Understanding Viscoelasticity

Advanced Texts in Physics

This program of advanced texts covers a broad spectrum of topics which are of current and emerging interest in physics. Each book provides a comprehensive and yet accessible introduction to a field at the forefront of modern research. As such, these texts are intended for senior undergraduate and graduate students at the MS and PhD level; however, research scientists seeking an introduction to particular areas of physics will also benefit from the titles in this collection.

Springer
Berlin
Heidelberg
New York
Barcelona
Hong Kong
London
Milan
Paris
Tokyo

Physics and Astronomy

ONLINE LIBRARY

http://www.springer.de/phys/

Nhan Phan-Thien

Understanding Viscoelasticity

Basics of Rheology

With 50 Figures

Springer

Professor Nhan Phan-Thien, FAA
Mechanical Engineering Department
National University of Singapore
Singapore 119260
Singapore

E-mail: nhan@nus.edu.sg

ISSN 1439-2674

ISBN 978-3-642-07779-1

Library of Congress Cataloging-in-Publication Data.
Phan-Thien, Nhan.
Understanding viscoelasticity: basics of rheology/Nhan Phan-Thien.
p.cm. – (Advanced texts in physics, ISSN 1439-2674)
Includes bibliographical references and index.

1. Viscoelasticity. I. Title. II. Series.
QA931.P426 2002   532'.0533–dc21   2002020926

This work is subject to copyright. All rights are reserved, whether the whole or part of the material is concerned, specifically the rights of translation, reprinting, reuse of illustrations, recitation, broadcasting, reproduction on microfilm or in any other way, and storage in data banks. Duplication of this publication or parts thereof is permitted only under the provisions of the German Copyright Law of September 9, 1965, in its current version, and permission for use must always be obtained from Springer-Verlag. Violations are liable for prosecution under the German Copyright Law.

Springer-Verlag Berlin Heidelberg New York
a member of BertelsmannSpringer Science+Business Media GmbH

http://www.springer.de

© Springer-Verlag Berlin Heidelberg 2010
Printed in Germany

The use of general descriptive names, registered names, trademarks, etc. in this publication does not imply, even in the absence of a specific statement, that such names are exempt from the relevant protective laws and regulations and therefore free for general use.

Cover design: *design & production* GmbH, Heidelberg
Printed on acid-free paper

Preface

This book presents an introduction to viscoelasticity; in particular, to the theories of dilute polymer solutions and melts, and dilute suspensions of rigid particles in viscous and incompressible fluids. These theories are important, not just because they apply to practical problems of industrial interest, but because they form a solid theoretical base upon which mathematical techniques can be built, from which more complex theories can be constructed, to better mimic material behaviour. The emphasis is not on the voluminous current topical research, but on the necessary tools to understand viscoelasticity at a first year graduate level.

Viscoelasticity, or Continuum Mechanics, or Rheology[1] (certainly not to be confused with Theology) is *the science of deformation and flow*. This definition of was due to Bingham, who, together with Scott-Blair[2] and Reiner,[3] helped form The Society of Rheology in 1929. Rheology has a distinguished history involving high-profile scientists. The idea that everything has a time scale, and that if we are prepared to wait long enough then everything will flow was known to the Greek philosopher Heraclitus, and prior to him, to the Prophetess Deborah – *The Mountains Flowed Before The Lord*.[4] Not surprisingly, the motto of the Society of Rheology is $\pi\alpha\nu\tau\alpha\,\rho\epsilon\iota$ (everything flows), a saying attributed to Heraclitus.

The logo of the Society of Rheology

From the rheological viewpoint, there is no clear distinction between solid and liquid, it is a matter between the time scale T of the experiment, and

---

[1] This word was coined by E.C. Bingham (1878–1946), Professor of Chemistry at Lafayette College, Pennsylvania. The Bingham fluid is named after him.

[2] G.W. Scott Blair (1902–1987), Professor of Chemistry at the University of Reading. His main contributions were in biorheology.

[3] M. Reiner (1886–1976), Professor of Mathematics at the Technion University of Haifa, Israel. He is remembered for contributing to the Reiner–Rivlin fluid.

[4] The Book of Judges.

the time scale τ of the material concerned. If the ratio is negligibly small, then one has a viscous fluid (more precise definition later), if it is large, a solid, and in-between, a viscoelastic liquid. The time scale ratio, $De = \tau/T$ is called the *Deborah number*. The time scale of the fluid varies considerably, from 10^{-13} s for water, to a few milliseconds for automotive oils, to minutes for polymer solutions, to hours for melts and soft solids.

Graduate students of Rheology naturally have the unenviable task of walking the bridge between solid mechanics and fluid mechanics, and at the same time trying to grasp the more significant and relevant concepts. They often find it hard (at least for me, during my graduate days) to piece together useful information from several comprehensive monographs and published articles on this subject. This set of lectures is an attempt to address this problem – it contains the necessary tools to understand viscoelasticity but does not insist on giving the latest piece of information on the topic.

The book starts with an introduction to the basic tools from tensor and dyadic analysis that are used in the book. Some authors prefer Cartesian tensor notation, others, dyadic notation. We use both notations and they will be summarised here. Chapter 2 is a review of non-Newtonian behaviour in flows; here the elasticity of the liquid and its ability to support large tension in stretching can be responsible for variety of phenomena, sometimes counter-intuitive. Kinematics and the equations of balance are discussed in details in Chapter 3, including the finite strain and Rivlin–Ericksen tensors. In Chapter 4 some classical constitutive equations are reviewed, and the general principles governing the constitutive modelling are outlined. In this Chapter, the order fluid models are also discussed, leading to the well-known result that the Newtonian velocity field is admissible to a second-order fluid in plane flow. Chapter 5 describes some of the popular engineering inelastic and the linear elastic models. The inelastic models are very useful in shear-like flows where viscosity/shear rate relation plays a dominant role. The linear viscoelastic model is a limit of the simple fluid at small strain – any model must reduce to this limit when the strain amplitude is small enough. In Chapter 6, we discuss a special class of flows known as viscometric flows in which both the kinematics and the stress are fully determined by the flow, irrespective of the constitutive equations. This class of flows is equivalent to the simple shearing flow. Modelling techniques for polymer solutions are discussed next in Chapter 7. Here one has a set of stochastic differential equations for the motion of the particles; the random excitations come from a white noise model of the collision between the solvent molecules and the particles. It is our belief that a relevant model should come from the microstructure; however, when the microstructure is so complex that a detailed model is not tractable, elements of continuum model should be brought in. Finally, an introduction to suspension mechanics is given in Chapter 8. I have deliberately left out a number of topics: instability, processing flows, electro-rheological fluids, magnitised fluids, and viscoelastic computational mechanics. It is hoped that the book forms a good foundation for those who wish to embark on the Rheology path.

This has been tested out in a one-semester course in Viscoelasticity at the National University of Singapore. It is entirely continuous-assessment based, with the assignments graded at different difficulty levels to be attempted – solving problems is an indispensable part of the education process. A good knowledge of fluid mechanics is helpful, but it is more important to have a solid foundation in Mathematics and Physics (Calculus, Linear Algebra, Partial Differential Equations), of a standard that every one gets in the first two years in an undergraduate Engineering curriculum.

I have greatly benefitted from numerous correspondence with my academic brother, Prof. Raj Huilgol and my mentor, Prof. Roger Tanner. Prof. Jeff Giacomin read the first draft of this; his help is gratefully acknowledged.

Singapore, February 2002   *Nhan Phan-Thien*

Contents

1. Tensor Notation

A working knowledge in tensor analysis

This chapter is not meant as a replacement for a course in tensor analysis, but it will provide a working background to tensor notation and algebra.

1.1 Cartesian Frame of Reference

Physical quantities encountered are either scalars (e.g., time, temperature, pressure, volume, density), or vectors (e.g., force, displacement, velocity, acceleration, force, torque, or tensors (e.g., stress, displacement gradient, velocity gradient, alternating tensors – we deal mostly with second-order tensors). These quantities are distinguished by the following generic notation:

    s denotes a scalar (lightface italic)
    u   denotes a vector (boldface)
    F   denotes a tensor (boldface)

The distinction between vector and tensor is usually clear from the context. When they are functions of points in a three-dimensional Euclidean space $\mathbb{E}$, they are called **fields**. The set of all vectors (or tensors) form a normed vector space $\mathbb{U}$.

Distances and time are measured in the Cartesian frame of reference, or simply frame of reference, $\mathcal{F} = \{O; \mathbf{e}_1, \mathbf{e}_2, \mathbf{e}_3\}$, which consists of an origin O, a clock, and an orthonormal basis $\{\mathbf{e}_1, \mathbf{e}_2, \mathbf{e}_3\}$, see Fig. 1.1,

$$\mathbf{e}_i \cdot \mathbf{e}_j = \delta_{ij}, \quad i, j = 1, 2, 3 \tag{1.1}$$

where the Kronecker delta is defined as

$$\delta_{ij} = \begin{cases} 1, & i = j, \\ 0, & i \neq j. \end{cases} \tag{1.2}$$

We only deal with right-handed frames of reference (applying the right-hand rule, when the thumb is in direction 1, and the forefinger in direction 2, the middle finger lies in direction 3), where $(\mathbf{e}_1 \times \mathbf{e}_2) \cdot \mathbf{e}_3 = 1$.

The Cartesian components of a vector **u** are given by

$$u_i = \mathbf{u} \cdot \mathbf{e}_i \tag{1.3}$$

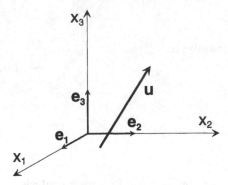

Fig. 1.1. Cartesian frame of reference

so that one may write

$$\mathbf{u} = \sum_{i=1}^{3} u_i \mathbf{e}_i = u_i \mathbf{e}_i. \tag{1.4}$$

Here we have employed the *summation convention*, i.e., whenever there are repeated subscripts, a summation is implied over the range of the subscripts, from (1, 2, 3). For example,

$$A_{ij}B_{jk} = \sum_{j=1}^{3} A_{ij}B_{jk}. \tag{1.5}$$

This short-hand notation is due to Einstein (Fig. 1.2), who argued that physical laws must not depend on coordinate systems, and therefore must be expressed in tensorial format. This is the essence of the *Principle of Frame Indifference*, to be discussed later.

Fig. 1.2. Albert Einstein (1879–1955) got the Nobel Prize in Physics in 1921 for his explanation in photoelectricity. He derived the effective viscosity of a dilute suspension of neutrally buoyant spheres, $\eta = \eta_s(1 + \frac{5}{2}\phi)$, η_s: the solvent viscosity, ϕ: the sphere volume fraction – in the original publication, the factor 5/2 was missing

The *alternating tensor* is defined as

$$\varepsilon_{ijk} = \begin{cases} +1 \text{ , if } (i,j,k) \text{ is an even permutation of } (1,2,3) \\ -1 \text{ , if } (i,j,k) \text{ is an odd permutation of } (1,2,3) \\ 0 \text{ , otherwise} \end{cases} \tag{1.6}$$

1.1.1 Position Vector

In the frame $\mathcal{F} = \{O; \mathbf{e}_1, \mathbf{e}_2, \mathbf{e}_3\}$, the position vector is denoted by

$$\mathbf{x} = x_i \mathbf{e}_i, \tag{1.7}$$

where x_i are the components of $\mathbf{x}$.

1.2 Frame Rotation

Consider the two frames of references, $\mathcal{F} = \{O; \mathbf{e}_1, \mathbf{e}_2, \mathbf{e}_3\}$ and $\mathcal{F}' = \{O; \mathbf{e}'_1, \mathbf{e}'_2, \mathbf{e}'_3\}$, as shown in Fig. 1.3, one obtained from the other by a rotation. Hence,

$$\mathbf{e}_i \cdot \mathbf{e}_j = \delta_{ij}, \quad \mathbf{e}'_i \cdot \mathbf{e}'_j = \delta_{ij}.$$

Define the cosine of the angle between $(\mathbf{e}_i, \mathbf{e}'_j)$ as

$$A_{ij} = \mathbf{e}'_i \cdot \mathbf{e}_j.$$

Thus A_{ij} can be regarded as the components of $\mathbf{e}'_i$ in $\mathcal{F}$, or the components of $\mathbf{e}_j$ in $\mathcal{F}'$. We write

$$\mathbf{e}'_p = A_{pi} \mathbf{e}_i \quad \therefore \quad A_{pi} A_{qi} = \delta_{pq}.$$

Similarly

$$\mathbf{e}_i = A_{pi} \mathbf{e}'_p \quad \therefore \quad A_{pi} A_{pj} = \delta_{ij}.$$

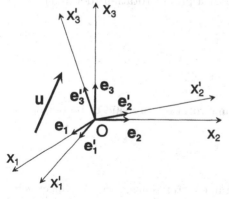

Fig. 1.3. Two frames of reference sharing a common origin

1.2.1 Orthogonal Matrix

A matrix is said to be an orthogonal matrix if its inverse is also its transpose; furthermore, if its determinant is $+1$, then it is a proper orthogonal matrix. Thus $[\mathbf{A}]$ is a proper orthogonal matrix.

We now consider a vector $|\mathbf{u}|$, with components

$$\mathbf{u} = u_i \mathbf{e}_i = u'_j \mathbf{e}'_j.$$

Taking scalar product with either base vector,

$$u'_i = \mathbf{e}'_i \cdot \mathbf{e}_j u_j = A_{ij} u_j,$$
$$u_j = \mathbf{e}_j \cdot \mathbf{e}'_i u_i = A_{ij} u'_i.$$

In matrix notation,

$$[\mathbf{A}] = \begin{bmatrix} A_{11} & A_{12} & A_{13} \\ A_{21} & A_{22} & A_{23} \\ A_{31} & A_{32} & A_{33} \end{bmatrix}, \quad [\mathbf{u}] = \begin{bmatrix} u_1 \\ u_2 \\ u_3 \end{bmatrix}, \quad [\mathbf{u}'] = \begin{bmatrix} u'_1 \\ u'_2 \\ u'_3 \end{bmatrix},$$

we have

$$[\mathbf{u}'] = [\mathbf{A}] \cdot [\mathbf{u}], \quad [\mathbf{u}] = [\mathbf{A}]^T \cdot [\mathbf{u}'],$$
$$u'_i = A_{ij} u_j, \qquad u_j = A_{ij} u'_i. \tag{1.8}$$

In particular, the position transforms according to this rule

$$\mathbf{x} = x'_i \mathbf{e}'_i = x_j \mathbf{e}_j \quad \therefore \quad x'_i = A_{ij} x_j \text{ or } x_j = A_{ij} x'_i.$$

1.2.2 Rotation Matrix

The matrix A is called a rotation – in fact a proper rotation ($\det = 1$).

1.3 Tensors

1.3.1 Zero-Order Tensors

Scalars, which are invariant under a frame rotation, are said to be tensors of zero order.

1.3.2 First-Order Tensor

A set of three scalars referred to one frame of reference, written collectively as $\mathbf{v} = (v_1, v_2, v_3)$, is called a tensor of first order, or a vector, if the three components transform according to (1.8) under a frame rotation.
Clearly,

- If $\mathbf{u}$ and $\mathbf{v}$ are vectors, then $\mathbf{u} + \mathbf{v}$ is also a vector.
- If $\mathbf{u}$ is a vector, then $\alpha\mathbf{u}$ is also a vector, where α is a real number.

The set of all vectors form a vector space $\mathcal{U}$ under addition and multiplication. In this space, the usual scalar product can be shown to be an inner product. With the norm induced by this inner product, $|\mathbf{u}|^2 = \mathbf{u} \cdot \mathbf{u}$, $\mathcal{U}$ is a normed vector space. We also refer to a vector $\mathbf{u}$ by its components, u_i.

1.3.3 Outer Products

Consider now two tensors of first order, u_i and v_i. The product $u_i v_j$ represents the outer product of $\mathbf{u}$ and $\mathbf{v}$, and written as (the subscripts are assigned from left to right by convention),

$$[\mathbf{uv}] = \begin{bmatrix} u_1 v_1 & u_1 v_2 & u_1 v_3 \\ u_2 v_1 & u_2 v_2 & u_2 v_3 \\ u_3 v_1 & u_3 v_2 & u_3 v_3 \end{bmatrix}.$$

In a frame rotation, from $\mathcal{F}$ to $\mathcal{F}'$, the components of this change according to

$$u_i' v_j' = A_{im} A_{jn} u_m u_n.$$

1.3.4 Second-Order Tensors

In general, a set of 9 scalars referred to one frame of reference, collectively written as $\mathbf{W} = [W_{ij}]$, transformed to another set under a frame rotation according to

$$W_{ij}' = A_{im} A_{jn} W_{mn}, \tag{1.9}$$

is said to be a second-order tensor, or a two-tensor, or simply a tensor (when the order does not have to be explicit). In matrix notation, we write

$$[\mathbf{W'}] = [\mathbf{A}]\,[\mathbf{W}]\,[\mathbf{A}]^T \text{ or } \mathbf{W'} = \mathbf{A}\mathbf{W}\mathbf{A}^T \text{ or } W_{ij}' = A_{ik} W_{kl} A_{jl}.$$

In the direct notation, we denote a tensor by a bold face letter (without the square brackets). This direct notation is intimately connected to the concept of a linear operator, e.g., Gurtin [1].

1.3.5 Third-Order Tensors

A set of 27 scalars referred to one frame of reference, collectively written as $\mathbf{W} = [W_{ijk}]$, transformed to another set under a frame rotation according to

$$W_{ijk}' = A_{il} A_{jm} A_{kn} W_{lmn}, \tag{1.10}$$

is said to be a third-order tensor.

Obviously, the definition can be extended to a set of 3^n scalars, and $\mathbf{W} = [W_{i_1 i_2 \ldots i_n}]$ (n indices) is said to be an n-order tensor if its components transform under a frame rotation according to

$$W'_{i_1 i_2 \ldots i_n} = A_{i_1 j_1} A_{i_2 j_2} \cdots A_{i_n j_n} W_{j_1 j_2 \ldots j_n}. \tag{1.11}$$

We will deal mainly with vector and tensors of second order. Usually, a higher-order (higher than 2) tensor is formed by taking outer products of tensors of lower orders, for example the outer product of a two-tensor $\mathbf{T}$ and a vector $\mathbf{n}$ is a third-order tensor $\mathbf{T} \otimes \mathbf{n}$. One can verify that the transformation rule (1.11) is obeyed.

1.3.6 Transpose Operation

The components of the transpose of a tensor $\mathbf{W}$ are obtained by swapping the indices:

$$[\mathbf{W}]_{ij} = W_{ij}, \quad [\mathbf{W}]_{ij}^T = W_{ji}.$$

A tensor $\mathbf{S}$ is *symmetric* if it is unaltered by the transpose operation,

$$\mathbf{S} = \mathbf{S}^T, \quad S_{ij} = S_{ji}.$$

It is *anti-symmetric* (or *skew*) if

$$\mathbf{S} = -\mathbf{S}^T, \quad S_{ij} = S_{ji}.$$

An anti-symmetric tensor must have zero diagonal terms (when $i = j$). Clearly

- If $\mathbf{U}$ and $\mathbf{V}$ are two-tensors, then $\mathbf{U} + \mathbf{V}$ is also a two-tensor.
- If $\mathbf{U}$ is a two-tensor, then $\alpha \mathbf{U}$ is also a two-tensor, where α is a real number. The set of $\mathbf{U}$ form a vector space under addition and multiplication.

1.3.7 Decomposition

Any second-order tensor can be decomposed into symmetric and anti-symmetric parts:

$$\mathbf{W} = \frac{1}{2} \left(\mathbf{W} + \mathbf{W}^T \right) + \frac{1}{2} \left(\mathbf{W} - \mathbf{W}^T \right), \tag{1.12}$$

$$W_{ij} = \frac{1}{2} \left(W_{ij} + W_{ji} \right) + \frac{1}{2} \left(W_{ij} - W_{ji} \right).$$

Returning to (1.9), if we interchange i and j, we get

$$W'_{ji} = A_{jm} A_{in} W_{mn} = A_{jn} A_{im} W_{nm}.$$

The second equality arises about because m and n are dummy indices, mere labels in the summation. The left side of this expression is recognized as the components of the transpose of $\mathbf{W}$. The equation asserts that the components of the transpose of $\mathbf{W}$ are also transformed according to (1.9). Thus, if $\mathbf{W}$ is a two-tensor, then its transpose is also a two-tensor, and the Cartesian decomposition (1.12) splits an arbitrary two-tensor into a symmetric and an anti-symmetric tensor (of second order).

We now go through some of the first and second-order tensors that will be encountered in this course.

1.3.8 Some Common Vectors

Position, displacement, velocity, acceleration, linear and angular momentum, linear and angular impulse, force, torque, are vectors. This is because the position vector transforms under a frame rotation according to (1.8). Any other quantity linearly related to the position (including the derivative and integral operation) will also be a vector.

1.3.9 Gradient of a Scalar

The gradient of a scalar is a vector. Let ϕ be a scalar, its gradient is written as

$$\mathbf{g} = \nabla\phi, \quad g_i = \frac{\partial\phi}{\partial x_i}.$$

Under a frame rotation, the new components of $\nabla\phi$ are

$$\frac{\partial\phi}{\partial x_i'} = \frac{\partial\phi}{\partial x_j}\frac{\partial x_j}{\partial x_i'} = A_{ij}\frac{\partial\phi}{\partial x_j},$$

which qualifies $\nabla\phi$ as a vector.

1.3.10 Some Common Tensors

We have met a second-order tensor formed by the outer product of two vectors, written compactly as $\mathbf{uv}$, with components (for vectors, the outer products is written without the symbol $\otimes$)

$$(\mathbf{uv})_{ij} = u_i v_j.$$

In general, the outer product of n vectors is an n-order tensor.

Unit Tensor. The Kronecker delta is a second-order tensor. In fact it is invariant in any coordinate system, and therefore is an *isotropic* tensor of second-order. To show that it is a tensor, note that

$$\delta_{ij} = A_{ik}A_{jk} = A_{ik}A_{jl}\delta_{kl},$$

which follows from the orthogonality of the transformation matrix. δ_{ij} are said to be the components of the second-order unit tensor $\mathbf{I}$. Finding isotropic tensors of arbitrary orders is not a trivial task.

Gradient of a Vector. The gradient of a vector is a two-tensor: if u_i and u_i' are the components of **u** in $\mathcal{F}$ and $\mathcal{F}'$,

$$\frac{\partial u_i'}{\partial x_j'} = \frac{\partial x_l}{\partial x_j'} \frac{\partial}{\partial x_l} \left(A_{ik} u_k \right) = A_{ik} A_{jl} \frac{\partial u_k}{\partial x_l}.$$

This qualifies the gradient of a vector as a two-tensor.

Velocity Gradient. If **u** is the velocity field, then ∇**u** is the gradient of the velocity. Be careful with the notation here. By our convention, the subscripts are assigned from left to right, so

$$(\nabla \mathbf{u})_{ij} = \nabla_i u_j = \frac{\partial u_j}{\partial x_i}.$$

In most books on viscoelasticity including this, the term velocity gradient is taken to mean the second-order tensor $\mathbf{L} = (\nabla \mathbf{u})^T$ with components

$$L_{ij} = \frac{\partial u_i}{\partial x_j}. \tag{1.13}$$

Strain Rate and Vorticity Tensors. The velocity gradient tensor can be decomposed into a symmetric part D, called the strain rate tensor, and an anti-symmetric part W, called the vorticity tensor:

$$\mathbf{D} = \frac{1}{2} \left(\nabla \mathbf{u} + \nabla \mathbf{u}^T \right), \quad \mathbf{W} = \frac{1}{2} \left(\nabla \mathbf{u}^T - \nabla \mathbf{u} \right). \tag{1.14}$$

Stress Tensor and Quotient Rule. We are given that stress $\mathbf{T} = [T_{ij}]$ at a point **x** is defined by, (see Fig. 1.4),

$$\mathbf{t} = \mathbf{Tn}, \quad t_i = T_{ij} n_j, \tag{1.15}$$

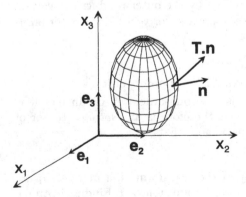

Fig. 1.4. Defining the stress tensor

where $\mathbf{n}$ is a normal unit vector on an infinitesimal surface ΔS at point $\mathbf{x}$, and $\mathbf{t}$ is the surface traction (force per unit area) representing the force the material on the positive side of $\mathbf{n}$ is pulling on the material on the negative side of $\mathbf{n}$. Under a frame rotation, since both $\mathbf{t}$ (force) and $\mathbf{n}$ are vectors,

$$\mathbf{t}' = \mathbf{At}, \ \ \mathbf{t} = \mathbf{A}^T\mathbf{t}' \qquad \mathbf{n}' = \mathbf{An}, \ \ \mathbf{n} = \mathbf{A}^T\mathbf{n}',$$
$$\mathbf{A}^T\mathbf{t}' = \mathbf{t} = \mathbf{Tn} = \mathbf{TA}^T\mathbf{n}' \ \ \therefore \ \ \mathbf{t}' = \mathbf{ATA}^T\mathbf{n}'.$$

From the definition of the stress, $\mathbf{t}' = \mathbf{T}'\mathbf{n}'$, and therefore

$$\mathbf{T}' = \mathbf{ATA}^T.$$

So the stress is a second-order tensor.

In fact, as long as $\mathbf{t}$ and $\mathbf{n}$ are vector, the 9 components T_{ij} defined in the manner indicated by (1.15) form a second-order tensor. This is known as the *quotient rule*.

1.4 Tensor and Linear Vector Function

L is a linear vector function on $\mathcal{U}$ if it satisfies

- $L(\mathbf{u}_1 + \mathbf{u}_2) = L(\mathbf{u}_1) + L(\mathbf{u}_2)$,
- $L(\alpha\mathbf{u}) = \alpha L(\mathbf{u}), \ \ \forall \mathbf{u}, \mathbf{u}_1, \mathbf{u}_2 \in \mathcal{U}, \ \ \forall \alpha \in \mathbb{R}$

1.4.1 Claim

Let $\mathbf{W}$ be a two-tensor, and define a vector-valued function through

$$\mathbf{v} = L(\mathbf{u}) = \mathbf{Wu},$$

then L is a linear function. Conversely, for any linear function on $\mathcal{U}$, there is a unique two-tensor $\mathbf{W}$ such that

$$L(\mathbf{u}) = \mathbf{Wu}, \ \ \forall \mathbf{u} \in \mathcal{U}.$$

The first statement can be easily verified. For the converse part, given the linear function, let define W_{ij} through

$$L(\mathbf{e}_i) = W_{ji}\mathbf{e}_j.$$

Now, $\forall \mathbf{u} \in \mathcal{U}$,

$$\mathbf{v} = L(\mathbf{u}) = L(u_i\mathbf{e}_i) = u_iW_{ji}\mathbf{e}_j$$
$$\therefore \ \ v_j = W_{ji}u_i.$$

$\mathbf{W}$ is a second-order tensor because $\mathbf{u}$ and $\mathbf{v}$ are vectors. The uniqueness part of $\mathbf{W}$ can be demonstrated by assuming that there is another $\mathbf{W}'$, then

$$(\mathbf{W} - \mathbf{W}')\,\mathbf{u} = 0, \quad \forall \mathbf{u} \in \mathcal{U},$$

which implies that $\mathbf{W}' = \mathbf{W}$.

In this connection, one can define a second-order tensor as a linear function, taking one vector into another. This is the direct approach, e.g., Gurtin [1], emphasising linear algebra. We use whatever notation is convenient for the purpose at hand. The set of all linear vector functions forms a vector space under addition and multiplication. The main result here is that

$$L\left(\mathbf{e}_i\right) = \mathbf{W}\mathbf{e}_i = W_{ji}\mathbf{e}_j \quad \therefore \quad W_{ji} = \mathbf{e}_j \cdot \left(\mathbf{W}\mathbf{e}_i\right).$$

1.4.2 Dyadic Notation

Thus, one may write

$$\mathbf{W} = W_{ij}\mathbf{e}_i\mathbf{e}_j. \tag{1.16}$$

This is the basis for the *dyadic* notation, the $\mathbf{e}_i\mathbf{e}_j$ play the role of the basis "vectors" for the tensor $\mathbf{W}$.

1.5 Tensor Operations

1.5.1 Substitution

The operation $\delta_{ij}u_j = u_i$ replaces the subscript j by i – the tensor δ_{ij} is therefore sometimes called the substitution tensor.

1.5.2 Contraction

Given a two-tensor W_{ij}, the operation

$$W_{ii} = \sum_{i=1}^{3} W_{ii} = W_{11} + W_{22} + W_{33}$$

is called a contraction. It produces a scalar. The invariance of this scalar under a frame rotation is seen by noting that

$$W'_{ii} = A_{ik}A_{il}W_{kl} = \delta_{kl}W_{kl} = W_{kk}.$$

This scalar is also called the trace of $\mathbf{W}$, written as

$$\text{tr}\,\mathbf{W} = W_{ii}. \tag{1.17}$$

It is one of the invariants of $\mathbf{W}$ (i.e., unchanged in a frame rotation). If the trace of $\mathbf{W}$ is zero, then $\mathbf{W}$ is said to be traceless. In general, given an n-order tensor, contracting any two subscripts produces a tensor of $(n-2)$ order.

1.5.3 Transpose

Given a two-tensor $\mathbf{W} = [W_{ij}]$, the transpose operation swaps the two indices

$$\mathbf{W}^T = (W_{ij}\mathbf{e}_i\mathbf{e}_j)^T = W_{ij}\mathbf{e}_j\mathbf{e}_i, \quad [\mathbf{W}^T]_{ij} = W_{ji}. \tag{1.18}$$

1.5.4 Products of Two Two-Tensors

Given two second-order tensors, $\mathbf{U}$ and $\mathbf{V}$,

$$\mathbf{U} = U_{ij}\mathbf{e}_i\mathbf{e}_j, \quad \mathbf{V} = V_{ij}\mathbf{e}_i\mathbf{e}_j,$$

one can form different products from them, and it is helpful to refer to the dyadic notation here.

- The tensor product $\mathbf{U} \otimes \mathbf{V}$ is a 4th-order tensor, with component $U_{ij}V_{kl}$,

$$\mathbf{U} \otimes \mathbf{V} = U_{ij}V_{kl}\mathbf{e}_i\mathbf{e}_j\mathbf{e}_k\mathbf{e}_l. \tag{1.19}$$

- The single dot product $\mathbf{U}.\mathbf{V}$ is a 2nd-order tensor, sometimes written without the dot (the dot is the contraction operator),

$$\begin{aligned}\mathbf{U} \cdot \mathbf{V} = \mathbf{U}\mathbf{V} &= (U_{ij}\mathbf{e}_i\mathbf{e}_j) \cdot (V_{kl}\mathbf{e}_k\mathbf{e}_l) \\ &= U_{ij}\mathbf{e}_i\delta_{jk}V_{kl}\mathbf{e}_l = U_{ij}V_{jl}\mathbf{e}_i\mathbf{e}_l,\end{aligned} \tag{1.20}$$

with components $U_{ik}V_{kl}$, just like multiplying two matrices U_{ik} and V_{kj}. This single dot product induces a contraction of a pair of subscripts (j and k) in $U_{ij}V_{kl}$, and acts just like a vector dot product.

- The double dot (or scalar, or inner) product produces a scalar,

$$\begin{aligned}\mathbf{U} : \mathbf{V} = (U_{ij}\mathbf{e}_i\mathbf{e}_j) : (V_{kl}\mathbf{e}_k\mathbf{e}_l) &= (U_{ij}\mathbf{e}_i)\,\delta_{jk} \cdot (V_{kl}\mathbf{e}_l) \\ &= U_{ij}V_{kl}\delta_{jk}\delta_{il} = U_{ij}V_{ji}.\end{aligned} \tag{1.21}$$

The dot operates on a pair of base vectors until we run out of dots. The end result is a scalar (remember our summation convention). It can be shown that the scalar product is in fact an inner product.

- The norm of a two-tensor is defined from the inner product in the usual manner,

$$\|\mathbf{U}\|^2 = \mathbf{U}^T : \mathbf{U} = U_{ij}U_{ij} = \mathrm{tr}\left(\mathbf{U}^T\mathbf{U}\right). \tag{1.22}$$

The space of all linear vector functions therefore form a normed vector space.

- One writes $\mathbf{U}^2 = \mathbf{U}\mathbf{U}$, $\mathbf{U}^3 = \mathbf{U}^2\mathbf{U}$, etc.
- A tensor $\mathbf{U}$ is invertible if there exists a tensor, $\mathbf{U}^{-1}$, called the inverse of $\mathbf{U}$, such that

$$\mathbf{U}\mathbf{U}^{-1} = \mathbf{U}^{-1}\mathbf{U} = \mathbf{I}. \tag{1.23}$$

One can also define the vector cross product between two two-tensors (and indeed any combination of dot and cross vector products). However, we refrain from listing all possible combinations here.

1.6 Invariants

1.6.1 Invariant of a Vector

When a quantity is unchanged with a frame rotation, it is said to be invariant. From a vector, a scalar can be formed by taking the scalar product with itself, $v_i v_i = v^2$. This is of course the magnitude of the vector and it is the only independent scalar invariant for a vector.

1.6.2 Invariants of a Tensor

From a second-order tensor $\mathbf{S}$, there are three independent scalar invariants that can be formed, by taking the trace of $\mathbf{S}$, $\mathbf{S}^2$ and $\mathbf{S}^3$,

$$I = \mathrm{tr}\mathbf{S} = S_{ii}, \quad II = \mathrm{tr}\mathbf{S}^2 = S_{ij}S_{ji}, \quad III = \mathrm{tr}\mathbf{S}^3 = S_{ij}S_{jk}S_{ki}.$$

However, it is customary to use the following invariants

$$I_1 = I, \quad I_2 = \frac{1}{2}\left(I^2 - II\right), \quad I_3 = \frac{1}{6}\left(I^3 - 3I\,II + 2III\right) = \det\mathbf{S}.$$

It is also possible to form ten invariants between two tensors (Gurtin [1]).

1.7 Decompositions

We now quote some of the well-known results without proof, some are intuitively obvious, others not.

1.7.1 Eigenvalue and Eigenvector

A scalar ω is an *eigenvalue* of a two-tensor $\mathbf{S}$ if there exists a non-zero vector $\mathbf{e}$, called the *eigenvector*, satisfying

$$\mathbf{Se} = \omega\mathbf{e}. \tag{1.24}$$

The characteristic space for $\mathbf{S}$ corresponding to the eigenvalue ω consists of all vectors in the eigenspace, $\{\mathbf{v} : \mathbf{Sv} = \omega\mathbf{v}\}$. If the dimension of this space is n, then ω is said to have geometric multiplicity of n. The *spectrum* of $\mathbf{S}$ is the ordered list $\{\omega_1, \omega_2, \ldots\}$ of all the eigenvalues of S.

A tensor $\mathbf{S}$ is said to be positive definite if it satisfies

$$\mathbf{S} : \mathbf{vv} > 0, \quad \forall\mathbf{v} \neq \mathbf{0}. \tag{1.25}$$

We record the following theorems:

- The eigenvalues of a positive definite tensor are strictly positive.

- The characteristic spaces of a symmetric tensor are mutually orthogonal.
- Spectral decomposition theorem: Let $\mathbf{S}$ be a symmetric two-tensor. Then there is a basis consisting entirely of eigenvectors of $\mathbf{S}$. For such a basis, $\{\mathbf{e}_i, i = 1, 2, 3\}$, the corresponding eigenvalues $\{\omega_i, i = 1, 2, 3\}$ form the entire spectrum of $\mathbf{S}$, and $\mathbf{S}$ can be represented by the *spectral representation*, where

$$
\begin{aligned}
&\mathbf{S} = \sum_{i=1}^{3} \omega_i \mathbf{e}_i \mathbf{e}_i, \text{ when } \mathbf{S} \text{ has three distinct eigenvalues,} \\
&\mathbf{S} = \omega_1 \mathbf{ee} + \omega_2 \left(\mathbf{I} - \mathbf{ee}\right), \text{ when } \mathbf{S} \text{ has two distinct eigenvalues,} \\
&\mathbf{S} = \omega \mathbf{I}, \text{ when } \mathbf{S} \text{ has only one eigenvalue.}
\end{aligned}
\tag{1.26}
$$

1.7.2 Square Root Theorem

Let $\mathbf{S}$ be a symmetric positive definite tensor. Then there is a unique positive definite tensor $\mathbf{U}$ such that $\mathbf{U}^2 = \mathbf{S}$. We write

$$
\mathbf{U} = \mathbf{S}^{1/2}.
$$

The proof of this follows from the spectral representation of $\mathbf{S}$.

1.7.3 Polar Decomposition Theorem

For any given tensor $\mathbf{F}$, there exist positive definite tensors $\mathbf{U}$ and $\mathbf{V}$, and a rotation tensor $\mathbf{R}$, such that

$$
\mathbf{F} = \mathbf{RU} = \mathbf{VR}.
\tag{1.27}
$$

Each of these representations is unique, and

$$
\mathbf{U} = \left(\mathbf{F}^T \mathbf{F}\right)^{1/2}, \quad \mathbf{V} = \left(\mathbf{FF}^T\right)^{1/2}.
\tag{1.28}
$$

The first representation ($\mathbf{RU}$) is called the right, and the second ($\mathbf{VR}$) is called the left polar decomposition.

1.7.4 Cayley–Hamilton Theorem

The most important theorem is the Cayley–Hamilton theorem: Every tensor $\mathbf{S}$ satisfies its own characteristic equation

$$
\mathbf{S}^3 - I_1 \mathbf{S}^2 + I_2 \mathbf{S} - I_3 \mathbf{I} = 0,
\tag{1.29}
$$

where $I_1 = \mathrm{tr}\mathbf{S}$, $I_2 = \frac{1}{2}\left((\mathrm{tr}\mathbf{S})^2 - \mathrm{tr}\mathbf{S}^2\right)$, and $I_3 = \det \mathbf{S}$ are the three scalar invariants for $\mathbf{S}$, and $\mathbf{I}$ is the unit tensor in three dimensions.

In two dimensions, this equation reads

$$\mathbf{S}^2 - I_1\mathbf{S} + I_2\mathbf{I} = 0, \tag{1.30}$$

where $I_1 = \mathrm{tr}\mathbf{S}$, $I_2 = \det\mathbf{S}$ are the two scalar invariants for $\mathbf{S}$, and $\mathbf{I}$ is the unit vector in two dimensions.

Cayley–Hamilton theorem is used to reduce the number of independent tensorial groups in tensor-valued functions. We record here one possible use of the Cayley–Hamilton theorem in two dimensions. The three-dimensional case is reserved as an exercise.

Suppose $\mathbf{C}$ is a given symmetric definite tensor in 2-D,

$$[\mathbf{C}] = \begin{bmatrix} C_{11} & C_{12} \\ C_{12} & C_{22} \end{bmatrix},$$

and its square root $\mathbf{U} = \mathbf{C}^{1/2}$ is desired. From the characteristic equation for $\mathbf{U}$,

$$\mathbf{U} = I_1^{-1}\left(\mathbf{U}\right)\left[\mathbf{C} + I_2\left(\mathbf{U}\right)\mathbf{I}\right],$$

so if we can express the invariants of $\mathbf{U}$ in terms of the invariant of $\mathbf{C}$, we're done. Now, if the eigenvalues of U are λ_1 and λ_2, then

$$I_1\left(\mathbf{U}\right) = \lambda_1 + \lambda_2, \quad I_2\left(\mathbf{U}\right) = \lambda_1\lambda_2,$$
$$I_1\left(\mathbf{C}\right) = \lambda_1^2 + \lambda_2^2, \quad I_2\left(\mathbf{C}\right) = \lambda_1^2\lambda_2^2.$$

Thus

$$I_2\left(\mathbf{U}\right) = \sqrt{I_2\left(\mathbf{C}\right)},$$
$$I_1^2\left(\mathbf{U}\right) = I_1\left(\mathbf{C}\right) + 2\sqrt{I_2\left(\mathbf{C}\right)}.$$

Therefore

$$\mathbf{U} = \frac{\mathbf{C} + \sqrt{I_2\left(\mathbf{C}\right)}\mathbf{I}}{\sqrt{I_1\left(\mathbf{C}\right) + 2\sqrt{I_2\left(\mathbf{C}\right)}}}.$$

1.8 Derivative Operations

We will not be overconcerned with the formal setting here. Suppose $\varphi\left(\mathbf{u}\right)$ is a scalar-valued function of a vector $\mathbf{u}$. The derivative of $\varphi(\mathbf{u})$ with respect to $\mathbf{u}$ in the direction $\mathbf{v}$ is defined as the linear operator $D\varphi\left(\mathbf{u}\right)\left[\mathbf{v}\right]$:

$$\varphi\left(\mathbf{u} + \alpha\mathbf{v}\right) = \varphi\left(\mathbf{u}\right) + \alpha D\varphi\left(\mathbf{u}\right)\left[\mathbf{v}\right] + HOT,$$

where HOT are terms of higher order, which vanish faster than α. Also, the square brackets enclosing $\mathbf{v}$ are used to emphasize the linearity of in $\mathbf{v}$. An operational definition for the derivative of $\varphi(\mathbf{u})$ in the direction $\mathbf{v}$ is therefore,

$$D\varphi\left(\mathbf{u}\right)\left[\mathbf{v}\right] = \frac{d}{d\alpha}\left[\varphi\left(\mathbf{u} + \alpha\mathbf{v}\right)\right]_{\alpha=0}. \tag{1.31}$$

This definition can be extended verbatim to derivatives of a tensor-valued (of any order) function of a tensor (of any order). The argument $\mathbf{v}$ is a part of the definition. We illustrate this with a few examples.

Example 1 Consider the scalar-valued function of a vector, $\varphi\left(\mathbf{u}\right)=u^2=\mathbf{u}\cdot\mathbf{u}$. Its derivative in the derection of $\mathbf{v}$ is

$$D\varphi\left(\mathbf{u}\right)\left[\mathbf{v}\right]=\frac{d}{d\alpha}\left[\mathbf{u}+\alpha\mathbf{v}\right]_{\alpha=0}=\frac{d}{d\alpha}\left[u^2+2\alpha\mathbf{u}\cdot\mathbf{v}+\alpha^2v^2\right]_{\alpha=0}$$

$$=2\mathbf{u}\cdot\mathbf{v}.$$

Example 2 Consider the tensor-valued function of a tensor, $\mathbf{G}\left(\mathbf{A}\right)=\mathbf{A}^2=\mathbf{A}\mathbf{A}$. Its derivative in the direction of $\mathbf{B}$ is

$$D\mathbf{G}\left(\mathbf{A}\right)\left[\mathbf{B}\right]=\frac{d}{d\alpha}\left[\mathbf{G}\left(\mathbf{A}+\alpha\mathbf{B}\right)\right]_{\alpha=0}$$

$$=\frac{d}{d\alpha}\left[\mathbf{A}^2+\alpha\left(\mathbf{A}\mathbf{B}+\mathbf{B}\mathbf{A}\right)+O\left(\alpha^2\right)\right]_{\alpha=0}$$

$$=\mathbf{A}\mathbf{B}+\mathbf{B}\mathbf{A}.$$

1.8.1 Derivative of det(A)

Consider the scalar-valued function of a tensor, $\varphi\left(\mathbf{A}\right)=\det\mathbf{A}$. Its derivative in the direction of $\mathbf{B}$ can be calculated using

$$\det(\mathbf{A}+\alpha\mathbf{B})=\det\alpha\mathbf{A}\left(\mathbf{A}^{-1}\mathbf{B}+\alpha^{-1}\mathbf{I}\right)=\alpha^3\det\mathbf{A}\det\left(\mathbf{A}^{-1}\mathbf{B}+\alpha^{-1}\mathbf{I}\right)$$

$$=\alpha^3\det\mathbf{A}\left(\alpha^{-3}+\alpha^{-2}I_1\left(\mathbf{A}^{-1}\mathbf{B}\right)+\alpha^{-1}I_2\left(\mathbf{A}^{-1}\cdot\mathbf{B}\right)+I_3\left(\mathbf{A}^{-1}\mathbf{B}\right)\right)$$

$$=\det\mathbf{A}\left(1+\alpha I_1\left(\mathbf{A}^{-1}\mathbf{B}\right)+O\left(\alpha^2\right)\right).$$

Thus

$$D\varphi\left(\mathbf{A}\right)\left[\mathbf{B}\right]=\frac{d}{d\alpha}\left[\varphi\left(\mathbf{A}+\alpha\mathbf{B}\right)\right]_{\alpha=0}=\det\mathbf{A}\mathrm{tr}\left(\mathbf{A}^{-1}\mathbf{B}\right).$$

1.8.2 Derivative of tr(A)

Consider the first invariant $I\left(\mathbf{A}\right)=\mathrm{tr}\mathbf{A}$. Its derivative in the direction of $\mathbf{B}$ is

$$DI\left(\mathbf{A}\right)\left[\mathbf{B}\right]=\frac{d}{d\alpha}\left[I\left(\mathbf{A}+\alpha\mathbf{B}\right)\right]_{\alpha=0}$$

$$=\frac{d}{d\alpha}\left[\mathrm{tr}\mathbf{A}+\alpha\mathrm{tr}\mathbf{B}\right]_{\alpha=0}=\mathrm{tr}\mathbf{B}=\mathbf{I}:\mathbf{B}.$$

1.8.3 Derivative of tr(A²)

Consider the second invariant $II\left(\mathbf{A}\right)=\mathrm{tr}\mathbf{A}^2$. Its derivative in the direction of $\mathbf{B}$ is

$$DII\left(\mathbf{A}\right)\left[\mathbf{B}\right]=\frac{d}{d\alpha}\left[II\left(\mathbf{A}+\alpha\mathbf{B}\right)\right]_{\alpha=0}$$

$$=\frac{d}{d\alpha}\left[\mathbf{A}:\mathbf{A}+\alpha\left(\mathbf{A}:\mathbf{B}+\mathbf{B}:\mathbf{A}\right)+O\left(\alpha^2\right)\right]_{\alpha=0}$$

$$=2\mathbf{A}:\mathbf{B}.$$

1.9 Gradient of a Field

1.9.1 Field

A function of the position vector $\mathbf{x}$ is called a field. One has a scalar field, for example the temperature field $T(\mathbf{x})$, a vector field, for example the velocity field $\mathbf{u}(\mathbf{x})$, or a tensor field, for example the stress field $\mathbf{S}(\mathbf{x})$. Higher-order tensor fields are rarely encountered, as in the many-point correlation fields. Conservation equations in continuum mechanics involve derivatives (derivatives with respect to position vectors are called *gradients*) of different fields, and it is absolutely essential to know how to calculate the gradients of fields in different coordinate systems. We also find it more convenient to employ the dyadic notation at this point.

1.9.2 Cartesian Frame

We consider first a scalar field, $\varphi(\mathbf{x})$. The Taylor expansion of this about point $\mathbf{x}$ is

$$\varphi(\mathbf{x} + \alpha \mathbf{r}) = \varphi(\mathbf{x}) + \alpha r_j \frac{\partial}{\partial x_j} \varphi(\mathbf{x}) + O(\alpha^2).$$

Thus the gradient of $\varphi(\mathbf{x})$ at point $\mathbf{x}$, now written as $\nabla \varphi$, defined in (1.31), is given by

$$\nabla \varphi[\mathbf{r}] = \mathbf{r} \cdot \frac{\partial \varphi}{\partial \mathbf{x}}. \qquad (1.32)$$

This remains unchanged for a vector or a tensor field.

Gradient Operator. This leads us to define the gradient operator as

$$\nabla = \mathbf{e}_j \frac{\partial}{\partial x_j} = \mathbf{e}_1 \frac{\partial}{\partial x_1} + \mathbf{e}_2 \frac{\partial}{\partial x_2} + \mathbf{e}_3 \frac{\partial}{\partial x_3}. \qquad (1.33)$$

This operator can be treated as a vector, operating on its arguments. By itself, it has no meaning; it must operate on a scalar, a vector or a tensor.

Gradient of a Scalar. For example, the gradient of a scalar is

$$\nabla \varphi = \mathbf{e}_j \frac{\partial \varphi}{\partial x_j} = \mathbf{e}_1 \frac{\partial \varphi}{\partial x_1} + \mathbf{e}_2 \frac{\partial \varphi}{\partial x_2} + \mathbf{e}_3 \frac{\partial \varphi}{\partial x_3}. \qquad (1.34)$$

Gradient of a Vector. The gradient of a vector can be likewise calculated

$$\nabla \mathbf{u} = \left(\mathbf{e}_i \frac{\partial}{\partial x_i} \right) (u_j \mathbf{e}_j) = \mathbf{e}_i \mathbf{e}_j \frac{\partial u_j}{\partial x_i}. \qquad (1.35)$$

In matrix notation,

$$[\nabla \mathbf{u}] = \begin{bmatrix} \dfrac{\partial u_1}{\partial x_1} & \dfrac{\partial u_2}{\partial x_1} & \dfrac{\partial u_3}{\partial x_1} \\[1.5ex] \dfrac{\partial u_1}{\partial x_2} & \dfrac{\partial u_2}{\partial x_2} & \dfrac{\partial u_3}{\partial x_2} \\[1.5ex] \dfrac{\partial u_1}{\partial x_3} & \dfrac{\partial u_2}{\partial x_3} & \dfrac{\partial u_3}{\partial x_3} \end{bmatrix}.$$

The component $(\nabla \mathbf{u})_{ij}$ is $\partial u_j / \partial x_i$; some books define this differently.

Transpose of a Gradient. The transpose of a gradient of a vector is therefore

$$\nabla \mathbf{u}^T = \mathbf{e}_i \mathbf{e}_j \frac{\partial u_i}{\partial x_j}. \tag{1.36}$$

In matrix notation,

$$[\nabla \mathbf{u}]^T = \begin{bmatrix} \dfrac{\partial u_1}{\partial x_1} & \dfrac{\partial u_1}{\partial x_2} & \dfrac{\partial u_1}{\partial x_3} \\[1.5ex] \dfrac{\partial u_2}{\partial x_1} & \dfrac{\partial u_2}{\partial x_2} & \dfrac{\partial u_2}{\partial x_3} \\[1.5ex] \dfrac{\partial u_3}{\partial x_1} & \dfrac{\partial u_3}{\partial x_2} & \dfrac{\partial u_3}{\partial x_3} \end{bmatrix}.$$

Divergence of a Vector. The divergence of a vector is a scalar defined by

$$\nabla \cdot \mathbf{u} = \left(\mathbf{e}_i \frac{\partial}{\partial x_i} \right) \cdot (u_j \mathbf{e}_j) = \mathbf{e}_i \cdot \mathbf{e}_j \frac{\partial u_j}{\partial x_i} = \delta_{ij} \frac{\partial u_j}{\partial x_i}$$

$$\nabla \cdot \mathbf{u} = \frac{\partial u_i}{\partial x_i} = \frac{\partial u_1}{\partial x_1} + \frac{\partial u_2}{\partial x_2} + \frac{\partial u_3}{\partial x_3}. \tag{1.37}$$

The divergence of a vector is also an invariant, being the trace of a tensor.

Curl of a Vector. The curl of a vector is a vector defined by

$$\nabla \times \mathbf{u} = \left(\mathbf{e}_i \frac{\partial}{\partial x_i} \right) \times (u_j \mathbf{e}_j) = \mathbf{e}_i \times \mathbf{e}_j \frac{\partial u_j}{\partial x_i} = \varepsilon_{kij} \mathbf{e}_k \frac{\partial u_j}{\partial x_i} \tag{1.38}$$

$$= \mathbf{e}_1 \left(\frac{\partial u_3}{\partial x_2} - \frac{\partial u_2}{\partial x_3} \right) + \mathbf{e}_2 \left(\frac{\partial u_1}{\partial x_3} - \frac{\partial u_3}{\partial x_1} \right) + \mathbf{e}_3 \left(\frac{\partial u_2}{\partial x_1} - \frac{\partial u_1}{\partial x_2} \right).$$

The curl of a vector is sometimes denoted by rot.

Divergence of a Tensor. The divergence of a tensor is a vector field defined by

$$\nabla \cdot \mathbf{S} = \left(\mathbf{e}_k \frac{\partial}{\partial x_k} \right) \cdot (S_{ij} \mathbf{e}_i \mathbf{e}_j) = \mathbf{e}_j \frac{\partial S_{ij}}{\partial x_i}. \tag{1.39}$$

1.9.3 Non-Cartesian Frames

All the above definition for gradient and divergence of a tensor remain valid in a non-Cartesian frame, provided that the derivative operation is also applied to the basis vectors as well. We illustrate this process in two important frames, cylindrical and spherical coordinate systems (Fig. 1.5); for other systems, consult Bird *et al.* [2].

Cylindrical Coordinates. In a cylindrical coordinate system (Fig. 1.5, left), points are located by giving them values to $\{r, \theta, z\}$, which are related to $\{x = x_1, y = x_2, z = x_3\}$ by

$$x = r\cos\theta, \quad y = r\sin\theta, \quad z = z$$
$$r = \sqrt{x^2 + y^2}, \quad \theta = \tan^{-1}\left(\frac{y}{x}\right), \quad z = z.$$

The basis vectors in this frame are related to the Cartesian ones by

$$\mathbf{e}_r = \cos\theta\mathbf{e}_x + \sin\theta\mathbf{e}_y, \quad \mathbf{e}_x = \cos\theta\mathbf{e}_r - \sin\theta\mathbf{e}_\theta$$
$$\mathbf{e}_\theta = -\sin\theta\mathbf{e}_x + \cos\theta\mathbf{e}_y, \quad \mathbf{e}_y = \sin\theta\mathbf{e}_r + \cos\theta\mathbf{e}_\theta.$$

Physical components. In this system, a vector $\mathbf{u}$, or a tensor $\mathbf{S}$, are represented by, respectively,

$$\mathbf{u} = u_r\mathbf{e}_r + u_\theta\mathbf{e}_\theta + u_z\mathbf{e}_z,$$
$$\mathbf{S} = S_{rr}\mathbf{e}_r\mathbf{e}_r + S_{r\theta}\mathbf{e}_r\mathbf{e}_\theta + S_{rz}\mathbf{e}_r\mathbf{e}_z + S_{\theta r}\mathbf{e}_\theta\mathbf{e}_r$$
$$+ S_{\theta\theta}\mathbf{e}_\theta\mathbf{e}_\theta + S_{\theta z}\mathbf{e}_\theta\mathbf{e}_z + S_{zr}\mathbf{e}_z\mathbf{e}_r + S_{z\theta}\mathbf{e}_z\mathbf{e}_\theta + S_{zz}\mathbf{e}_z\mathbf{e}_z.$$

Gradient operator. The components expressed this way are called physical components. The gradient operator is converted from one system to another by the chain rule,

$$\nabla = \mathbf{e}_x\frac{\partial}{\partial x} + \mathbf{e}_y\frac{\partial}{\partial y} + \mathbf{e}_z\frac{\partial}{\partial z} = (\cos\theta\mathbf{e}_r - \sin\theta\mathbf{e}_\theta)\left(\cos\theta\frac{\partial}{\partial r} - \frac{\sin\theta}{r}\frac{\partial}{\partial\theta}\right)$$
$$+ (\sin\theta\mathbf{e}_r + \cos\theta\mathbf{e}_\theta)\left(\sin\theta\frac{\partial}{\partial r} + \frac{\cos\theta}{r}\frac{\partial}{\partial\theta}\right) + \mathbf{e}_z\frac{\partial}{\partial z}$$
$$= \mathbf{e}_r\frac{\partial}{\partial r} + \mathbf{e}_\theta\frac{1}{r}\frac{\partial}{\partial\theta} + \mathbf{e}_z\frac{\partial}{\partial z}. \tag{1.40}$$

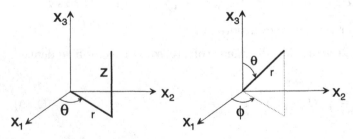

Fig. 1.5. Cylindrical and spherical frame of references

When carrying out derivative operations, remember that

$$\frac{\partial}{\partial r}\mathbf{e}_r = 0, \quad \frac{\partial}{\partial r}\mathbf{e}_\theta = 0, \quad \frac{\partial}{\partial r}\mathbf{e}_z = 0$$

$$\frac{\partial}{\partial \theta}\mathbf{e}_r = \mathbf{e}_\theta, \quad \frac{\partial}{\partial \theta}\mathbf{e}_\theta = -\mathbf{e}_r, \quad \frac{\partial}{\partial \theta}\mathbf{e}_z = 0 \qquad (1.41)$$

$$\frac{\partial}{\partial z}\mathbf{e}_r = 0, \quad \frac{\partial}{\partial z}\mathbf{e}_\theta = 0, \quad \frac{\partial}{\partial z}\mathbf{e}_z = 0.$$

Gradient of a vector. The gradient of any vector is

$$\nabla \mathbf{u} = \left(\mathbf{e}_r \frac{\partial}{\partial r} + \mathbf{e}_\theta \frac{1}{r}\frac{\partial}{\partial \theta} + \mathbf{e}_z \frac{\partial}{\partial z} \right) (u_r \mathbf{e}_r + u_\theta \mathbf{e}_\theta + u_z \mathbf{e}_z)$$

$$= \mathbf{e}_r\mathbf{e}_r \frac{\partial u_r}{\partial r} + \mathbf{e}_r\mathbf{e}_\theta \frac{\partial u_\theta}{\partial r} + \mathbf{e}_r\mathbf{e}_z \frac{\partial u_z}{\partial r} + \mathbf{e}_\theta\mathbf{e}_r \frac{1}{r}\frac{\partial u_r}{\partial \theta} + \mathbf{e}_\theta\mathbf{e}_\theta \frac{u_r}{r}$$

$$+ \mathbf{e}_\theta\mathbf{e}_\theta \frac{1}{r}\frac{\partial u_\theta}{\partial \theta} - \mathbf{e}_\theta\mathbf{e}_r \frac{u_\theta}{r} + \mathbf{e}_\theta\mathbf{e}_z \frac{1}{r}\frac{\partial u_z}{\partial \theta} + \mathbf{e}_z\mathbf{e}_r \frac{\partial u_r}{\partial z} + \mathbf{e}_z\mathbf{e}_\theta \frac{\partial u_\theta}{\partial z}$$

$$+ \mathbf{e}_z\mathbf{e}_z \frac{\partial u_z}{\partial z}$$

$$\nabla \mathbf{u} = \mathbf{e}_r\mathbf{e}_r \frac{\partial u_r}{\partial r} + \mathbf{e}_r\mathbf{e}_\theta \frac{\partial u_\theta}{\partial r} + \mathbf{e}_r\mathbf{e}_z \frac{\partial u_z}{\partial r} + \mathbf{e}_\theta\mathbf{e}_r \left(\frac{1}{r}\frac{\partial u_r}{\partial \theta} - \frac{u_\theta}{r} \right)$$

$$+ \mathbf{e}_\theta\mathbf{e}_\theta \left(\frac{1}{r}\frac{\partial u_\theta}{\partial \theta} + \frac{u_r}{r} \right) + \mathbf{e}_\theta\mathbf{e}_z \frac{1}{r}\frac{\partial u_z}{\partial \theta} + \mathbf{e}_z\mathbf{e}_r \frac{\partial u_r}{\partial z} + \mathbf{e}_z\mathbf{e}_\theta \frac{\partial u_\theta}{\partial z}$$

$$+ \mathbf{e}_z\mathbf{e}_z \frac{\partial u_z}{\partial z}. \qquad (1.42)$$

Divergence of a vector. The divergence of a vector is obtained by a contraction of the above equation:

$$\nabla \cdot \mathbf{u} = \frac{\partial u_r}{\partial r} + \frac{1}{r}\frac{\partial u_\theta}{\partial \theta} + \frac{u_r}{r} + \frac{\partial u_z}{\partial z}. \qquad (1.43)$$

1.9.4 Spherical Coordinates

In a spherical coordinate system (Fig. 1.5, right), points are located by giving them values to $\{r, \theta, \phi\}$, which are related to $\{x = x_1, y = x_2, z = x_3\}$ by

$$x = r \sin\theta \cos\phi, \quad y = r \sin\theta \sin\phi, \quad z = r \cos\theta,$$

$$r = \sqrt{x^2 + y^2 + z^2}, \quad \theta = \tan^{-1}\left(\frac{\sqrt{x^2 + y^2}}{z} \right), \quad \phi = \tan^{-1}\left(\frac{y}{x} \right).$$

The basis vectors are related by

$$\mathbf{e}_r = \mathbf{e}_1 \sin\theta \cos\phi + \mathbf{e}_2 \sin\theta \sin\phi + \mathbf{e}_3 \cos\theta,$$

$$\mathbf{e}_\theta = \mathbf{e}_1 \cos\theta \cos\phi + \mathbf{e}_2 \cos\theta \sin\phi - \mathbf{e}_3 \sin\theta,$$

$$\mathbf{e}_\phi = -\mathbf{e}_1 \sin\phi + \mathbf{e}_2 \cos\phi,$$

and

$$e_1 = e_r \sin\theta \cos\phi + e_\theta \cos\theta \cos\phi - e_\phi \sin\phi,$$
$$e_2 = e_r \sin\theta \sin\phi + e_\theta \cos\theta \sin\phi + e_\phi \cos\phi,$$
$$e_3 = e_r \cos\theta - e_\theta \sin\theta.$$

Gradient operator. Using the chain rule, it can be shown that the gradient operator in spherical coordinates is

$$\nabla = e_r \frac{\partial}{\partial r} + e_\theta \frac{1}{r}\frac{\partial}{\partial \theta} + e_\phi \frac{1}{r\sin\theta}\frac{\partial}{\partial \phi}. \tag{1.44}$$

We list below a few results of interest.
Gradient of a scalar. The gradient of a scalar is given by

$$\nabla\varphi = e_r \frac{\partial\varphi}{\partial r} + e_\theta \frac{1}{r}\frac{\partial\varphi}{\partial \theta} + e_\phi \frac{1}{r\sin\theta}\frac{\partial\varphi}{\partial \phi}. \tag{1.45}$$

Gradient of a vector. The gradient of a vector is given by

$$\nabla u = e_r e_r \frac{\partial u_r}{\partial r} + e_r e_\theta \frac{\partial u_\theta}{\partial r} + e_r e_\phi \frac{\partial u_\phi}{\partial r} + e_\theta e_r \left(\frac{1}{r}\frac{\partial u_r}{\partial \theta} - \frac{u_\theta}{r}\right)$$
$$+ e_\theta e_\theta \left(\frac{1}{r}\frac{\partial u_\theta}{\partial \theta} + \frac{u_r}{r}\right) + e_\phi e_r \left(\frac{1}{r\sin\theta}\frac{\partial u_r}{\partial \phi} - \frac{u_\phi}{r}\right)$$
$$+ e_\theta e_\phi \frac{1}{r}\frac{\partial u_\phi}{\partial \theta} + e_\phi e_\theta \left(\frac{1}{r\sin\theta}\frac{\partial u_\theta}{\partial \phi} - \frac{u_\phi}{r}\cot\theta\right)$$
$$+ e_\phi e_\phi \left(\frac{1}{r\sin\theta}\frac{\partial u_\phi}{\partial \phi} + \frac{u_r}{r} + \frac{u_\theta}{r}\cot\theta\right). \tag{1.46}$$

Divergence of a vector. The divergence of a vector is given by

$$\nabla \cdot u = \frac{1}{r^2}\frac{\partial}{\partial r}\left(r^2 u_r\right) + \frac{1}{r}\frac{\partial}{\partial \theta}\left(u_\theta \sin\theta\right) + \frac{1}{r\sin\theta}\frac{\partial u_\phi}{\partial \phi}. \tag{1.47}$$

Divergence of a tensor. The divergence of a tensor is given by

$$\nabla \cdot S = e_r \left[\frac{1}{r^2}\frac{\partial}{\partial r}\left(r^2 S_{rr}\right) + \frac{1}{r\sin\theta}\frac{\partial}{\partial \theta}\left(S_{\theta r}\sin\theta\right) + \frac{1}{r\sin\theta}\frac{\partial S_{\phi r}}{\partial \phi}\right. \tag{1.48}$$
$$\left. - \frac{S_{\theta\theta} + S_{\phi\phi}}{r}\right] + e_\theta \left[\frac{1}{r^3}\frac{\partial}{\partial r}\left(r^3 S_{r\theta}\right) + \frac{1}{r\sin\theta}\frac{\partial}{\partial \theta}\left(S_{\theta\theta}\sin\theta\right)\right.$$
$$\left. + \frac{1}{r\sin\theta}\frac{\partial S_{\phi\theta}}{\partial \phi} + \frac{S_{\theta r} - S_{r\theta} - S_{\phi\phi}\cot\theta}{r}\right] + e_\phi \left[\frac{1}{r^3}\frac{\partial}{\partial r}\left(r^3 S_{r\phi}\right)\right.$$
$$\left. + \frac{1}{r\sin\theta}\frac{\partial}{\partial \theta}\left(S_{\theta\phi}\sin\theta\right) + \frac{1}{r\sin\theta}\frac{\partial S_{\phi\phi}}{\partial \phi} + \frac{S_{\phi r} - S_{r\phi} + S_{\phi\theta}\cot\theta}{r}\right].$$

1.10 Integral Theorems

1.10.1 Gauss Divergence Theorem

Various volume integrals can be converted to surface integrals by the following theorems, due to Gauss (Fig. 1.6):

$$\int_V \nabla \varphi dV = \int_S \varphi \mathbf{n} dS, \tag{1.49}$$

$$\int_V \nabla \cdot \mathbf{u} dV = \int_S \mathbf{n} \cdot \mathbf{u} dS, \tag{1.50}$$

$$\int_V \nabla \cdot \mathbf{S} dV = \int_S \mathbf{n} \cdot \mathbf{S} dS. \tag{1.51}$$

The proofs may be found in Kellogg [3]. In these, V is a bounded regular region (Fig. 1.7), and φ (or $\mathbf{u}$ or $\mathbf{S}$) are differentiable scalar (or vector or tensor) field with continuous gradients. Indeed the indicial version of (1.9) is valid even if u_i are merely three scalar fields of the required smoothness (rather than three components of a vector field).

Fig. 1.6. Carl Friedrich Gauss (1777–1855) was a Professor of Mathematics at the University of Göttingen until he died. He made several contributions to Number Theory, Geodesy, Statistics, Geometry, Physics. His motto was few, but ripe (*Pauca, sed matura*), and nothing further remains to be done. He did not publish several important papers because they did not satisfy these requirements

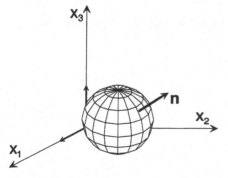

Fig. 1.7. A region enclosed by a closed surface with outward unit vector field

1.10.2 Stokes Curl Theorem

Various surfaces integrals can be converted into contour integrals using the following theorems:

$$\int_S \mathbf{n} \cdot (\nabla \times \mathbf{u})\, dS = \oint_C \mathbf{t} \cdot \mathbf{u}\, dC, \tag{1.52}$$

$$\int_S \mathbf{n} \cdot (\nabla \times \mathbf{S})\, dS = \oint_C \mathbf{t} \cdot \mathbf{S}\, dC. \tag{1.53}$$

In these, $\mathbf{t}$ is a tangential unit vector along the contour C. The direction of integration is determined by the right-hand rule: thumb pointing in the direction of $\mathbf{n}$, fingers curling in the direction of C.

1.10.3 Leibniz Formula

If φ is a field (a scalar, a vector, or a tensor) define on a region $V(t)$, which is changing in time, with bounding surface $S(t)$, also changing in time with velocity $\mathbf{u}_S$, then

$$\frac{d}{dt}\int_V \varphi\, dV = \int_V \frac{\partial \varphi}{\partial t}\, dV + \int_S \varphi \mathbf{u}_S \cdot \mathbf{n}\, dS. \tag{1.54}$$

Fig. 1.8. Gottfried W. Leibniz (1646–1716) was a German philosopher and mathematician, who independently with Newton, laid the foundation for integral and differential calculus in 1675

Problems

Problem 1.1 The components of vectors $\mathbf{u}$, $\mathbf{v}$, and $\mathbf{w}$ are given by u_i, v_i, w_i. Verify that

$$\mathbf{u} \cdot \mathbf{v} = u_i v_i,$$

$$\mathbf{u} \times \mathbf{v} = \varepsilon_{ijk} \mathbf{e}_i u_j v_k,$$

$$(\mathbf{u} \times \mathbf{v}) \cdot \mathbf{w} = \varepsilon_{ijk} u_i v_j w_k,$$

$$(\mathbf{u} \times \mathbf{v}) \cdot \mathbf{w} = \mathbf{u} \cdot (\mathbf{v} \times \mathbf{w}),$$

$$(\mathbf{u} \times \mathbf{v}) \times \mathbf{w} = (\mathbf{u} \cdot \mathbf{w})\, \mathbf{v} - (\mathbf{v} \cdot \mathbf{w})\, \mathbf{u},$$

$$(\mathbf{u} \times \mathbf{v})^2 = u^2 v^2 - (\mathbf{u} \cdot \mathbf{v})^2,$$

where $u^2 = |\mathbf{u}|^2$ and $v^2 = |\mathbf{v}|^2$.

Problem 1.2 Let $\mathbf{A}$ be a 3×3 matrix with entries A_{ij},

$$[\mathbf{A}] = \begin{bmatrix} A_{11} & A_{12} & A_{13} \\ A_{21} & A_{22} & A_{23} \\ A_{31} & A_{32} & A_{33} \end{bmatrix}.$$

Verify that

$$\det[\mathbf{A}] = \varepsilon_{ijk} A_{1i} A_{2j} A_{3k} = \varepsilon_{ijk} A_{1i} A_{j2} A_{k3},$$
$$\varepsilon_{lmn} \det[\mathbf{A}] = \varepsilon_{ijk} A_{il} A_{jm} A_{kn} = \varepsilon_{ijk} A_{li} A_{mj} A_{nk},$$
$$\det[\mathbf{A}] = \frac{1}{6}\varepsilon_{ijk}\varepsilon_{lmn} A_{il} A_{jm} A_{kn}.$$

Problem 1.3 Verify that

$$\varepsilon_{ijk}\varepsilon_{imn} = \delta_{jm}\delta_{kn} - \delta_{jn}\delta_{km}.$$

Given that two 3×3 matrices of components

$$[\mathbf{A}] = \begin{bmatrix} A_{11} & A_{12} & A_{13} \\ A_{21} & A_{22} & A_{23} \\ A_{31} & A_{32} & A_{33} \end{bmatrix}, \quad [\mathbf{B}] = \begin{bmatrix} B_{11} & B_{12} & B_{13} \\ B_{21} & B_{22} & B_{23} \\ B_{31} & B_{32} & B_{33} \end{bmatrix}$$

verify that if $[\mathbf{C}] = [\mathbf{A}] \cdot [\mathbf{B}]$, then the components of $\mathbf{C}$ are $C_{ij} = A_{ik}B_{kj}$. Thus if $[\mathbf{D}] = [\mathbf{A}]^T [\mathbf{B}]$, then $D_{ij} = A_{ki}B_{kj}$.

Problem 1.4 Show that

$$\det[A_{ij}] = (\mathbf{e}'_1 \times \mathbf{e}'_2) \cdot \mathbf{e}'_3 = 1.$$
$$[\mathbf{A}]^T [\mathbf{A}] = [\mathbf{A}][\mathbf{A}]^T = [\mathbf{I}], \quad [\mathbf{A}]^{-1} = [\mathbf{A}]^T, \quad \det[\mathbf{A}] = 1.$$

Problem 1.5 Verify that

$$\varepsilon_{ijk}u_i v_j w_k = \det \begin{bmatrix} u_1 & u_2 & u_3 \\ v_1 & v_2 & v_3 \\ w_1 & w_2 & w_3 \end{bmatrix}.$$

Consider a two-tensor W_{ij} and a vector $u_i = \varepsilon_{ijk}W_{jk}$. Show that if $\mathbf{W}$ is symmetric, $\mathbf{u}$ is zero, and if $\mathbf{W}$ is anti-symmetric the components of $\mathbf{u}$ are twice those of $\mathbf{W}$. This vector is said to be the axial vector of $\mathbf{W}$.
Hence, show that the axial vector associated with the vorticity tensor of (1.14) is $-\nabla \times \mathbf{u}$.

Problem 1.6 If $\mathbf{D}$, $\mathbf{S}$, and $\mathbf{W}$ are two-tensors, $\mathbf{D}$ symmetric and $\mathbf{W}$ anti-symmetric, show that

$$\mathbf{D} : \mathbf{S} = \mathbf{D} : \mathbf{S}^T = \mathbf{D} : \tfrac{1}{2}\left(\mathbf{S} + \mathbf{S}^T\right),$$
$$\mathbf{W} : \mathbf{S} = -\mathbf{W} : \mathbf{S}^T = \mathbf{W} : \tfrac{1}{2}\left(\mathbf{W} - \mathbf{W}^T\right),$$
$$\mathbf{D} : \mathbf{W} = 0.$$

Further, show that

if $\mathbf{T}:\mathbf{S}=0$ $\forall \mathbf{S}$ then $\mathbf{T}=0$,

if $\mathbf{T}:\mathbf{S}=0$ $\forall$ symmetric $\mathbf{S}$ then $\mathbf{T}$ is anti-symmetric,

if $\mathbf{T}:\mathbf{S}=0$ $\forall$ anti-symmetric $\mathbf{S}$ then $\mathbf{T}$ is symmetric.

Problem 1.7 Show that $\mathbf{Q}$ is orthogonal if and only if $\mathbf{H}=\mathbf{Q}-\mathbf{I}$ satisfies

$$\mathbf{H}+\mathbf{H}^T+\mathbf{H}\mathbf{H}^T=0, \quad \mathbf{H}\mathbf{H}^T=\mathbf{H}^T\mathbf{H}.$$

Problem 1.8 Show that I, II, III are indeed invariants. In addition, show that

$$\det(\mathbf{S}-\omega\mathbf{I})=-\omega^3+I_1\omega^2-I_2\omega+I_3.$$

If ω is an eigenvalue of $\mathbf{S}$ then $\det(\mathbf{S}-\omega\mathbf{I})=0$. This is said to be the characteristic equation for $\mathbf{S}$.

Problem 1.9 Apply the result above to find the square root of the Cauchy-Green tensor in a shear deformation

$$[\mathbf{C}]=\begin{bmatrix} 1+\gamma^2 & \gamma \\ \gamma & 1 \end{bmatrix}.$$

Investigate the corresponding formula for the square root of a symmetric positive definite tensor $\mathbf{S}$ in three dimensions.

Problem 1.10 Write down all the components of the strain rate tensor and the vorticity tensor.

Problem 1.11 Given that $\mathbf{r}=x_i\mathbf{e}_i$ is the position vector, $\mathbf{a}$ is a constant vector, and $f(r)$ is a function of $r=|\mathbf{r}|$, show that

$$\nabla\cdot\mathbf{r}=3, \quad \nabla\times\mathbf{r}=\mathbf{0}, \quad \nabla(\mathbf{a}\cdot\mathbf{r})=\mathbf{a}, \quad \nabla f=\frac{1}{r}\frac{df}{dr}\mathbf{r}.$$

Problem 1.12 Show that the divergence of a tensor in cylindrical coordinates is given by

$$\begin{aligned}
\nabla\cdot\mathbf{S}=\mathbf{e}_r &\left(\frac{\partial S_{rr}}{\partial r}+\frac{S_{rr}-S_{\theta\theta}}{r}+\frac{1}{r}\frac{\partial S_{\theta r}}{\partial\theta}+\frac{\partial S_{zr}}{\partial z}\right) \\
+\mathbf{e}_\theta &\left(\frac{\partial S_{r\theta}}{\partial r}+\frac{2S_{r\theta}}{r}+\frac{1}{r}\frac{\partial S_{\theta\theta}}{\partial\theta}+\frac{\partial S_{z\theta}}{\partial z}+\frac{S_{\theta r}-S_{r\theta}}{r}\right) \\
+\mathbf{e}_z &\left(\frac{\partial S_{rz}}{\partial r}+\frac{S_{rz}}{r}+\frac{1}{r}\frac{\partial S_{\theta z}}{\partial\theta}+\frac{\partial S_{zz}}{\partial z}\right).
\end{aligned} \tag{1.55}$$

Problem 1.13 Show that

$$\begin{aligned}
\nabla^2\mathbf{u}=\mathbf{e}_r &\left[\frac{\partial}{\partial r}\left(\frac{1}{r}\frac{\partial}{\partial r}(ru_r)\right)+\frac{1}{r^2}\frac{\partial^2 u_r}{\partial\theta^2}+\frac{\partial^2 u_r}{\partial z^2}-\frac{2}{r^2}\frac{\partial u_\theta}{\partial\theta}\right] \\
+\mathbf{e}_\theta &\left[\frac{\partial}{\partial r}\left(\frac{1}{r}\frac{\partial}{\partial r}(ru_\theta)\right)+\frac{1}{r^2}\frac{\partial^2 u_\theta}{\partial\theta^2}+\frac{\partial^2 u_\theta}{\partial z^2}+\frac{2}{r^2}\frac{\partial u_r}{\partial\theta}\right] \\
+\mathbf{e}_z &\left[\frac{1}{r}\frac{\partial}{\partial r}\left(r\frac{\partial u_z}{\partial r}\right)+\frac{1}{r^2}\frac{\partial^2 u_z}{\partial\theta^2}+\frac{\partial^2 u_z}{\partial z^2}\right].
\end{aligned} \tag{1.56}$$

Problem 1.14 Show that

$$
\begin{aligned}
\mathbf{u} \cdot \nabla \mathbf{u} = \mathbf{e}_r & \left[u_r \frac{\partial u_r}{\partial r} + \frac{u_\theta}{r} \frac{\partial u_r}{\partial \theta} + u_z \frac{\partial u_r}{\partial z} - \frac{u_\theta u_\theta}{r} \right] \\
+ \mathbf{e}_\theta & \left[u_r \frac{\partial u_\theta}{\partial r} + \frac{u_\theta}{r} \frac{\partial u_\theta}{\partial \theta} + u_z \frac{\partial u_\theta}{\partial z} + \frac{u_\theta u_r}{r} \right] \\
+ \mathbf{e}_z & \left[u_r \frac{\partial u_z}{\partial r} + \frac{u_\theta}{r} \frac{\partial u_z}{\partial \theta} + u_z \frac{\partial u_z}{\partial z} \right].
\end{aligned}
\tag{1.57}
$$

Problem 1.15 The stress tensor in a material satisfies $\nabla \cdot \mathbf{S} = \mathbf{0}$. Show that the volume-average stress in a region V occupied by the material is

$$
\langle \mathbf{S} \rangle = \frac{1}{2V} \int_S (\mathbf{xt} + \mathbf{tx}) \, dS,
\tag{1.58}
$$

where $\mathbf{t} = \mathbf{n} \cdot \mathbf{S}$ is the surface traction. The quantity on the left side of (1.58) is called the *stresslet* (Batchelor [4]).

Problem 1.16 Calculate the following integrals on the surface of the unit sphere

$$
\langle \mathbf{nn} \rangle = \frac{1}{S} \int_S \mathbf{nn} \, dS
\tag{1.59}
$$

$$
\langle \mathbf{nnnn} \rangle = \frac{1}{S} \int_S \mathbf{nnnn} \, dS.
\tag{1.60}
$$

These are the averages of various moments of a uniformly distributed unit vector on a sphere surface.

Problem 1.?. Show that

2. Rheological Properties

Overall material properties and flow behaviour

Fluids with featureless microstructures are well described by the Newtonian constitutive equation, which states that the stress tensor is proportional to the shear rate tensor (these concepts will be made precise later). Fluids with complex microstructures, for example suspensions of particles or droplets (blood, paint, ink, asphalt, bitumen, foodstuffs, etc.), polymer melts and solutions (molten plastics, fibre-reinforced or particulate plastics), exhibit a wide variety of behaviours. These are summarised here. For more information, consult Bird *et al.* [2] and Tanner [5].

2.1 Viscosity

2.1.1 Shear-Rate Dependent Viscosity

The most important for engineering calculation is the fluid viscosity. This quantity is defined as the ratio of the shear stress to the shear rate.

Here, as shown in Fig. 2.1, the flow is generated by sliding one plate atop another, with the fluid in-between. The quantities of interest are the shear rate, $\dot{\gamma} = U/h$ (U is the velocity of the top plate, h is the sample thickness), $S = F/A$ is the shear stress (F is the tangential force on the top plate, A the fluid contact area). The shear stress is an odd function of the shear rate. In addition, with viscoelastic fluids, there may be a normal force on the plates.

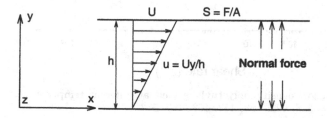

Fig. 2.1. Shear flow generated by sliding one plate on top of another. Shear force as well as normal force may be supplied to keep the plates at a fixed distance

When a steady flow can be established, the viscosity is defined as

$$\eta = \frac{S}{\dot{\gamma}}. \tag{2.1}$$

For a Newtonian fluid, this is a constant (having units Pa.s) depending only on the temperature. For most fluids with long chain microstructure (polymer melts and solutions), the viscosity is a decreasing function of the shear rate, sometimes reaching of the zero-shear rate viscosity. This type of behaviour is called *shear thinning*. The opposite behaviour, *shear thickening*, is sometimes observed with some suspensions, due to the formation of clusters. A typical viscosity-shear rate curve is shown in Fig. 2.2 for a low-density polyethylene (LDPE) at different temperatures.

It can be seen that the viscosity is a strong function of the temperature; a 60°C increase in the temperature induces a ten-fold decrease in viscosity. In addition, the viscosity decreases by an order of magnitude (compared to its zero-shear-rate value) at a shear rate of $\dot{\gamma} = 10^{-1}$ s^{-1}. A constant viscosity does not qualify a fluid to be Newtonian. The term Newtonian is much more restrictive in its meaning.

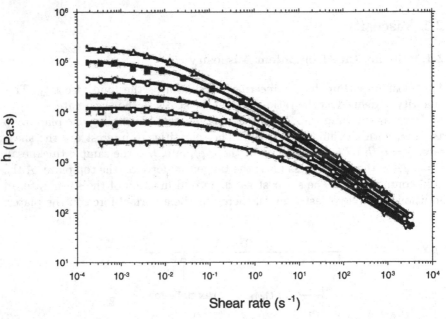

Fig. 2.2. Viscosity of a low density polyethylene melt at different temperature. From top to bottom, $T = 388, 403, 423, 443, 463, 483, 513$ K

For some materials with solid-like behaviour (for example, bread dough, biological tissues), viscosity measurement makes no sense, since the shear stress just keeps increasing with time until the sample breaks or flows out of the test cell, and what has been measured is not a material property, but an indication of the friction between the sample and the test apparatus.

With suspensions of particles with surface charges, one can get viscosity to behave in many different ways; even a discontinuity at a particular shear rate may be induced.

Incidentally, the CGS units for viscosity is Poise, in honour of Poiseuille (Fig. 2.3), who provided the experimental results for pipe flow in 1846; (1 Pa.s = 10 Poises). The viscosity of water is about 1 cP (centi-Poises).

Fig. 2.3. Jean-Louis-Marie Poiseuille (1799–1869) was a French physisian who estalished experimentally the pressure-drop/flow rate relationship of laminar flow in tubes

2.2 Normal Stress Differences

Normal stress differences refer to the differences between the unequal normal stresses in shear flow (for a Newtonian fluid in shear flow, the normal stresses are always equal). With three normal stress components, we can form two, the *first* and the *second normal stress differences*

$$N_1 = S_{xx} - S_{yy}, \quad N_2 = S_{yy} - S_{zz}, \tag{2.2}$$

These normal stress differences are even functions of the shear rate, and therefore one defines the normal stress coefficients as

$$\nu_1 = \frac{N_1}{\dot{\gamma}^2}, \quad \nu_2 = \frac{N_2}{\dot{\gamma}^2}. \tag{2.3}$$

These normal stress coefficients are even functions of the shear rate. The normal stress differences and the shear viscosity are called viscometric functions; they are the material functions representing the viscometric properties of the fluid.

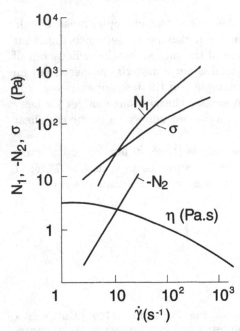

Fig. 2.4. Viscometric functions of 6.8% of polyisobutylene in cetane at 24°C

Figure 2.4 shows some typical measurements of viscometric properties of a polyisobutylene solution, the first (N_1) and the second (N_2) normal stress differences, the shear stress (σ) and the viscosity.

The second normal stress difference is not usually measured. In general, it is negative, and about 10% of N_1 in magnitude.

Suspensions have non-zero normal stress differences as well; however, our knowledge of them is still incomplete. Non-equal normal stresses are responsible for some visually striking differences between Newtonian and non-Newtonian fluid. We summarise the key features here.

2.2.1 Weissenberg Rod-Climbing Effect

When a rod is rotating in viscoelastic fluid, the fluid climbs the rod rather than depressing the free surface near the rod. This phenomenon is called the Weissenberg[1] rod-climbing effect. Rod climbing is due to the fluid element being able to support a tension along a streamline (first normal stress difference) and thus it forces the fluid up the rod. This effect can occur without the rod: if one rotates a disk at the bottom of the beaker, then the middle of the free surface on the fluid bulges.

---

[1] Karl Weissenberg (1893–1976) contributed significantly to Rheology in the early years, and has several phenomema named after him.

Fig. 2.5. Weissenberg rod climbing effect

2.2.2 Die Swell

When a viscoelastic fluid exits from a capillary of diameter D, it tends to swell considerably more than a Newtonian fluid. For highly viscous Newtonian fluids, the swell ratio, D_E/D, where D_E is the extrudate diameter, is at most 13%. For a polymer melt, the extrudate diameter could be a few times the capillary diameter. This phenomenon is called *die swell*, and the dominant mechanism causing this is the first normal stress difference. In fact, Tanner [6] proposed the simple rule for capillary die swell, based on a simple analysis

$$\frac{D_E}{D} = 0.13 + \left[1 + \frac{1}{2}\left(\frac{N_1}{2S}\right)_w^2 \right]^{1/6}, \tag{2.4}$$

where N_1 and S are the first normal stress difference and the shear stress, both evaluated at the wall (subscript w). Die swell is mainly due to the fluid elasticity, but it can also occur with the shear thinning induced by viscous

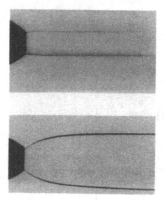

Fig. 2.6. Die swell – top: a Newtonian fluid, bottom: a viscoelastic fluid

heating. Inertia tends to reduce the amount of swell, and to delay it, see Fig. 2.7.

Fig. 2.7. Delay die swell – increasing Reynolds number from left to right

2.2.3 Flow Down an Inclined Channel

The second normal stress difference, although small in magnitude, is important in some cases. In the flow down an inclined channel, a Newtonian fluid is seen to have a nearly flat free surface, whereas a convex surface is seen for a viscoelastic fluid with a negative second normal stress difference.

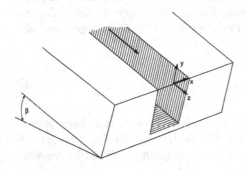

Fig. 2.8. In a flow down an inclined channel, the free surface will bulge up if N_2 is negative

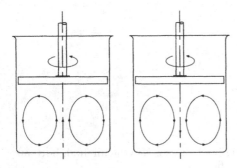

Fig. 2.9. Reversed secondary flow – left: Newtonian fluid, right: viscoelastic fluid

Viscoelasticity is also responsible for the reversal of the secondary flow pattern; one such case is sketched in Fig. 2.9.

2.3 Transient Responses

Viscoelastic fluids have a relaxation time scale, and this can be quantified in several ways.

2.3.1 Small Strain Oscillatory Flow

In an oscillatory shear flow (Fig. 2.1), where the top plate oscillates sinusoidally with angular frequency ω, $x = \delta \sin \omega t$, the plate velocity is $U = \omega \delta \cos \omega t$. The shear rate and the shear strain are given by

$$\dot{\gamma} = \dot{\gamma}_0 \cos \omega t, \quad \gamma = \gamma_0 \sin \omega t, \tag{2.5}$$
$$\dot{\gamma}_0 = \delta \omega / h, \quad \gamma_0 = \delta / h.$$

When the strain is small, the shear stress is also sinusoidal, but is not in phase with either the strain or the strain rate,

$$S = S' \sin \omega t + S'' \cos \omega t.$$

The part that is in phase with the strain is used to define the storage modulus, or the storage viscosity

$$G' = \frac{S'}{\gamma_0}, \quad \eta'' = \frac{S'}{\dot{\gamma}_0}, \quad G' = \omega \eta'', \tag{2.6}$$

and the part that is in phase with the strain rate is used to define the loss modulus, or the dynamic viscosity,

$$G'' = \frac{S''}{\gamma_0}, \quad \eta' = \frac{S''}{\dot{\gamma}_0}, \quad G'' = \omega \eta'. \tag{2.7}$$

These are functions of the frequency, and they are termed *dynamic properties*.

Figure 2.10 show the storage and loss moduli of LDPE at different temperatures. The data have been collapsed into a master curve through the use of the time-temperature superposition principle, which involves scaling the frequency by a *shift factor* a_T. The dynamic properties contain time scale information on the fluid expressed in the frequency domain. Large-amplitude oscillatory tests have also been done, but their interpretation is less straightforward.

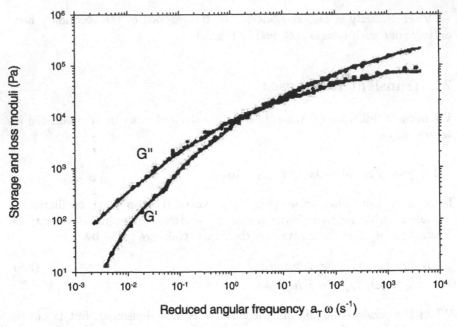

Fig. 2.10. Dynamic properties of a low density polyethylene (LDPE) melt

2.3.2 Stress Overshoot

In a start-up of a shear flow, i.e., $U = U_0 H(t)$, or $\dot{\gamma} = \dot{\gamma}_0 H(t)$, where $H(t)$ is the Heaviside function, the shear stress increases with time, then overshoots before approaching its steady value, sometimes with a few oscillations about the steady value. This is seen with the normal stress differences as well. One can define the unsteady viscosity, like the steady-state viscosity:

$$\eta^+ (t, \dot{\gamma}_0) = \frac{S(t)}{\dot{\gamma}_0}. \tag{2.8}$$

The amount of stress overshoot can be significant at high shear rates, and it has been speculated that this is why certain biological fluid (e.g., synovial fluid) is a good lubricant.

2.3.3 Stress Relaxation

Corresponding to the start-up of shear flow is stress relaxation, where the fluid motion that has been undergoing a steady-state shear flow at a shear rate is suddenly stopped. The shear stress (and the normal stress differences) is monitored as it relaxes. Again, one can define the stress-relaxation viscosity

$$\eta^- (t, \dot{\gamma}_0) = \frac{S(t)}{\dot{\gamma}_0}. \tag{2.9}$$

2.3.4 Relaxation Modulus

There is another type of relaxation experiment, by applying a large strain rate over a small interval, so that the total strain is $\gamma_0 = \dot{\gamma}_0 \Delta t$, and monitoring the shear (and normal) stress as it relaxes. This allows the *relaxation modulus* to be defined:

$$G(t, \gamma_0) = \frac{S(t, \gamma_0)}{\gamma_0}. \qquad (2.10)$$

At small enough strains, $G(t, \gamma_0) = G_0(t)$ is independent of the strain, because the stress is linear in the strain at low strains. The relaxation modulus of LDPE is shown in Fig. 2.11. The parallelism of the curves (at different strains) suggests that the relaxation modulus can be factored in a function of strain and a function of time,

$$G(t, \gamma_0) = h(\gamma_0)G_0(t), \quad h(0) = 1. \qquad (2.11)$$

This is known as *strain-time separability*.

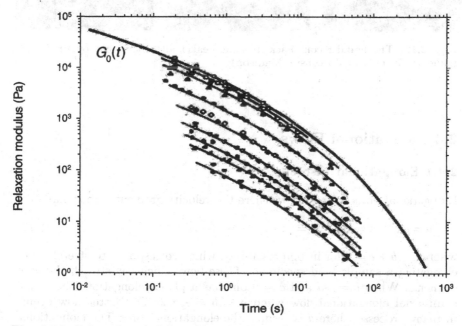

Fig. 2.11. Relaxation modulus of low density polyethylene (LDPE) – increasing strain from top to bottom

2.3.5 Recoil

If the loading is suddenly removed, as seen in Fig. 2.12 by cutting the liquid column, the liquid retracts to some previous shape. The liquid is said to have memory (it remembers where it came from). However, its memory is imperfect, as it can only retract partially. In that sense it has fading memory. A Newtonian liquid has a catastrophic memory: the moment the loading is removed, the motion ceases immediately. An elastic solid has a perfect memory: upon removal of the loads, the solid particles return to exactly the positions they occupy previously.

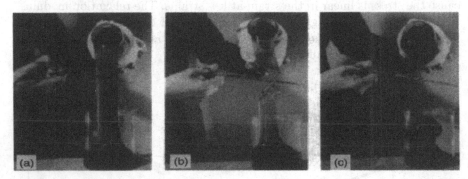

Fig. 2.12. The liquid recoils back into the beaker after being cut by Prof. A.S. Lodge (University of Wisconsin, Madison)

2.4 Elongational Flows

2.4.1 Elongational Viscosity

Elongational flows refer to flow where the velocity gradient is diagonal, i.e.,

$$u = ax, \quad v = by, \quad w = cz, \tag{2.12}$$

where $a + b + c = 0$ for incompressibility, which corresponds to stretching or elongating a sample fluid specimen. These flows efficiently stretch the fluid elements. When $b = -a$ and $c = 0$ one has a planar elongational flow, and a uni-axial elongational flow when $a = b = -c/2$. This latter flow occurs in many processes; here a is termed the elongational rate. The elongational viscosity is defined as

$$\eta_E = \frac{S_{xx} - S_{yy}}{a}. \tag{2.13}$$

Except at very low elongational rates, elongational viscosity does not usually reach a steady state (the sample elongates and fails). For a Newtonian fluid its elongational viscosity is thrice its shear viscosity, but for a polymer solution, the elongational viscosity can be orders of magnitude greater. The Trouton ratio is define to the ratio of the eleongation viscosity to the shear viscosity of the fluid

$$\text{Trouton Ratio} = \frac{\eta_E}{\eta} \qquad\qquad (2.14)$$

A typical plot of the Trouton ratio for a polybutene solution is shown in Fig. 2.13. For a Newtonian fluid, the Trouton ratio is three, for a viscoelastic fluid, this ratio may be very large.

The ability of a liquid filament to support a significant tensile stress is mainly why the tubeless siphon experiment (Fig. 2.14) works.

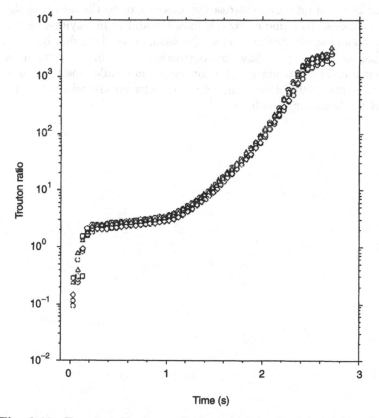

Fig. 2.13. Transient Trouton ratio for a high molecular weight polyisobutylene solution – the extensional rate is $a = 2\ \text{s}^{-1}$

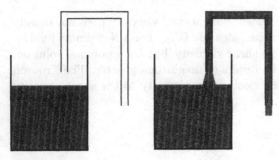

Fig. 2.14. Tubeless siphon

2.5 Viscoelastic Instabilities

Because of the non-linearity in the constitutive equations, viscoelastic flows are full of instabilities. These instabilities may not depend on inertia; they are mainly driven by the fluid normal stresses (elasticity), or by the nature of the boundary conditions. To name a few, we have instability in Taylor-Couette flow, in the torsional flow between two parallel disks, in the shear flow between cone-and-plate, in curved pipe flow, in contraction flows, in the flows from extrusion dies, etc. The extrudate distortion, commonly called melt fracture, is an example of instability due to the interplay between viscoelasticity and the nature of the boundary conditions.

3. Kinematics and Equations of Balance

A quick review of Continuum Mechanics

In this chapter, we review the kinematics and the equations of balance (conservation equations), leaving the question of constitutive description to the next chapter. Huilgol & Phan-Thien [7] provides additional reading material.

3.1 Kinematics

3.1.1 Reference Configuration

We deal with a continuous body $\mathcal{B}$, which occupies a region consisting of points in $\mathbb{E}_3$. We refer to one particular configuration, $\mathcal{B}_R$, for example at time $t = 0$, as the reference configuration. The particle position in the reference configuration is denoted by a capital letter $\mathbf{X}$. This particle traces out a path in $\mathbb{E}_3$ and its current position is denoted by the small letter $\mathbf{x}$; of course $\mathbf{x}$ is a function of time. The particle is referred to by its position in the reference configuration.

Fig. 3.1. Joseph-Louis Lagrange (1736–1813) was a Italian/French mathematician, who made important contributions to analytic mechanics, the calculus of variations, and number theory. His book *Mécanique Analytique* was an approved publication by a committee including Laplace and Legendre in 1788

A motion is defined to be a twice-differentiable and invertible map (so that acceleration field can be defined, and that every position $\mathbf{x}$ must correspond to a particle $\mathbf{X}$)

$$\mathbf{x} = \mathbf{M}(\mathbf{X}, t), \quad x_i = M_i(X_j, t), \tag{3.1}$$

where t is the time. This is also called the *Lagrangian description* (see Fig. 3.1) of the motion. Note that $\mathbf{X} = \mathbf{M}(\mathbf{X}, 0)$. Collectively $\mathbf{M}(\mathbf{X}, t)$, $\mathbf{X} \in \mathcal{B}_R$

gives us the spatial description of the motion, called the *current configuration*. Since $\mathbf{M}$ is invertible,

$$\mathbf{X} = \mathbf{M}^{-1}(\mathbf{x}, t), \quad X_i = M_i^{-1}(x_j, t) \tag{3.2}$$

gives us the reference in terms of the current configuration.

3.1.2 Velocity and Acceleration Fields

The velocity and the acceleration fields are defined as

$$\hat{\mathbf{u}} = \frac{\partial}{\partial t}\mathbf{M}(\mathbf{X}, t), \quad \hat{u}_i = \frac{\partial}{\partial t}M_i(X_j, t), \tag{3.3}$$

and

$$\hat{\mathbf{a}} = \frac{\partial^2}{\partial t^2}\mathbf{M}(\mathbf{X}, t), \quad \hat{a}_i = \frac{\partial^2}{\partial t^2}M_i(X_j, t), \tag{3.4}$$

respectively.

It is customary to refer to velocity and acceleration fields as functions of the current position, the so-called *Eulerian description* (see Fig. 3.2). This is accomplished using (3.2).

Fig. 3.2. Leonhard Euler (1707–1783) was one of the greatest mathematicians of the 18th century. He occupied Daniel Bernouilli's chair of mathematics at St. Petersburg. He perfected the integral calculus, worked in analytic geometry, theory of lunar motion, introduced the Euler's identity and many other mathematical symbols that we are familiar with, π, $f()$, $\sum$. He published more than 856 articles and books

Writing as the Eulerian velocity field, we find that the Eulerian acceleration field is given by

$$
\begin{aligned}
\mathbf{a} &= \left.\frac{\partial}{\partial t}\mathbf{u}(\mathbf{x}, t)\right|_{\mathbf{x}} = \frac{\partial\mathbf{u}}{\partial t} + \left.\frac{\partial\mathbf{u}}{\partial\mathbf{x}} \cdot \frac{\partial\mathbf{x}}{\partial t}\right|_{\mathbf{x}} \\
&= \frac{\partial\mathbf{u}}{\partial t} + \mathbf{u} \cdot \nabla\mathbf{u} = \frac{\partial\mathbf{u}}{\partial t} + \mathbf{L} \cdot \mathbf{u},
\end{aligned} \tag{3.5}
$$

where we have introduce the velocity gradient tensor

$$\mathbf{L} = \left(\frac{\partial\mathbf{u}}{\partial\mathbf{x}}\right)^T = (\nabla\mathbf{u})^T, \quad L_{ij} = \frac{\partial u_i}{\partial x_j}. \tag{3.6}$$

Our convention assigns subscripts from left to right. An Eulerian velocity is called steady if it does not depend on time, i.e., $\mathbf{u} = \mathbf{u}(\mathbf{x})$. A steady Eulerian velocity field is thus not necessarily Lagrangian steady.

3.1.3 Material Derivative

The derivative on a spatial field embodied in (3.5) is called the material derivative,

$$\frac{d}{dt} = \frac{\partial}{\partial t} + \mathbf{u} \cdot \nabla \qquad (3.7)$$

The symmetric part of the velocity gradient is called the strain rate tensor $\mathbf{D}$, and its anti-symmetric part is called the vorticity tensor,

$$\mathbf{L} = \mathbf{D} + \mathbf{W}, \quad \mathbf{D} = \frac{1}{2}\left(\mathbf{L} + \mathbf{L}^T\right), \quad \mathbf{W} = \frac{1}{2}\left(\mathbf{L} - \mathbf{L}^T\right). \qquad (3.8)$$

3.2 Deformation Gradient and Strain Tensors

3.2.1 Deformation Gradient

The gradient of $\mathbf{x}$ with respect to $\mathbf{X}$ is called the deformation gradient,

$$\mathbf{F} = \left(\frac{\partial \mathbf{x}}{\partial \mathbf{X}}\right)^T, \quad F_{ij} = \frac{\partial x_i}{\partial X_j}. \qquad (3.9)$$

Note again our subscript convention. At time $t = 0$ the initial value of $\mathbf{F}$ is

$$\mathbf{F}(0) = \mathbf{I}, \qquad (3.10)$$

the identity tensor. The mass in a region V is

$$\int_V \rho dx = \int_{V_0} \rho |J| \, d\mathbf{X},$$

where $J = \det \mathbf{F}$ and V_0 is the region occupied by the reference configuration. Thus, we demand that $\det \mathbf{F} > 0$, so that the mapping is not degenerate. For an incompressible fluid, the kinematic constraint is of course

$$\det \mathbf{F} = 1. \qquad (3.11)$$

Because of the chain rule,

$$\left(\frac{\partial \mathbf{X}}{\partial \mathbf{x}}\right)^T = \mathbf{F}^{-1}, \quad F_{ij}^{-1} = \frac{\partial X_i}{\partial x_j}. \qquad (3.12)$$

The connection between the deformation and the velocity gradients arises from the equality

$$\frac{\partial}{\partial X_j}\left(\frac{\partial M_i}{\partial t}\right) = \frac{\partial}{\partial t}\left(\frac{\partial M_i}{\partial X_j}\right) \quad \frac{\partial \hat{u}_i}{\partial X_j} = \frac{\partial F_{ij}}{\partial t}.$$

Using the Eulerian description for the velocity

$$\frac{\partial \hat{u}_i}{\partial X_j} = \frac{\partial u_i}{\partial x_k}\frac{\partial x_k}{\partial X_j} = L_{ik}F_{kj},$$

one has

$$\dot{\mathbf{F}} = \mathbf{L}\mathbf{F}, \quad \mathbf{F}(0) = \mathbf{I}, \tag{3.13}$$

where the super dot denotes the time derivative. This equation provides an initial-value problem for $\mathbf{F}$.

3.2.2 Cauchy–Green Strain Tensor

The concept of strain is introduced by comparing the length of a fluid element at the current time to that in the reference configuration. We have from the definition of the deformation gradient,

$$d\mathbf{x} = \mathbf{F}d\mathbf{X}.$$

Here $d\mathbf{X}$ is a fluid element at point $\mathbf{X}$, which at time t has changed to $d\mathbf{x}$ at point $\mathbf{x}$, see Fig. 3.3. Its current length is

$$dx^2 = d\mathbf{x}\cdot d\mathbf{x} = F_{ij}dX_j F_{ik}dX_k = d\mathbf{X}\mathbf{F}^T \cdot \mathbf{F}d\mathbf{X} = \mathbf{F}^T\mathbf{F} : d\mathbf{X}d\mathbf{X}$$

The tensor

$$\mathbf{C} = \mathbf{F}^T\mathbf{F} \tag{3.14}$$

is therefore a measure of the strain the fluid experiences, it is called the *right Cauchy–Green* tensor (for a portray of Cauchy, see Fig. 3.4). The *left Cauchy–Green* tensor is defined as

$$\mathbf{B} = \mathbf{F}\mathbf{F}^T. \tag{3.15}$$

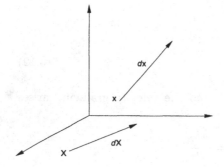

Fig. 3.3. An element $d\mathbf{X}$ at $\mathbf{X}$ in the reference configuration at time $t = 0$ is changed to $d\mathbf{x}$ at $\mathbf{x}$ at time t

Fig. 3.4. Augustin Cauchy (1789–1857) was a prolific French mathematician. He contributed more than 16 fundamental concepts and theorems, and published more than 800 papers. His name is one of the 72 inscribed on the Eiffel tower

The name refers to the right or left polar decompositions of $\mathbf{F}$,

$$\mathbf{F} = \mathbf{RU} \text{ (right)} = \mathbf{VR} \text{ (left)},$$

where $\mathbf{U}$ and $\mathbf{V}$ are symmetric positive-definite tensors (right and left stretch tensors), and $\mathbf{R}$ is an orthogonal tensor. Thus

$$\mathbf{C} = \mathbf{F}^T\mathbf{F} = \mathbf{U}^2, \quad \mathbf{B} = \mathbf{FF}^T = \mathbf{V}^2.$$

The inverse of the Cauchy–Green tensor is also used; it is called the Finger strain tensor.

3.2.3 Relative Strain Tensors

The reference configuration enjoys no particular mathematical status, although it may have a physical significance (e.g., the stress-free state, for example). Suppose the particle X at time τ occupies the position ξ. The relative deformation gradient is defined as

$$\mathbf{F}_t(\tau) = \left(\frac{\partial \xi}{\partial \mathbf{x}}\right)^T, \quad (\mathbf{F}_t(\tau))_{ij} = \frac{\partial \xi_i}{\partial x_j}, \tag{3.16}$$

and correspondingly, the right (or left) relative Cauchy–Green tensor

$$\mathbf{C}_t(\tau) = \mathbf{F}_t(\tau)^T \mathbf{F}_t(\tau), \quad (\mathbf{C}_t(\tau))_{ij} = \frac{\partial \xi_k}{\partial x_i}\frac{\partial \xi_k}{\partial x_j}. \tag{3.17}$$

Because of the chain rule,

$$\mathbf{F}_t(\tau) = \left(\frac{\partial \xi}{\partial \mathbf{x}}\right)^T = \left(\frac{\partial \mathbf{X}}{\partial \mathbf{x}} \cdot \frac{\partial \xi}{\partial \mathbf{X}}\right)^T = \left(\frac{\partial \xi}{\partial \mathbf{X}}\right)^T \cdot \left(\frac{\partial \mathbf{X}}{\partial \mathbf{x}}\right)^T$$

$$= \mathbf{F}(\tau)\mathbf{F}(t)^{-1}. \tag{3.18}$$

3.2.4 Path Lines

To solve for the path lines of the particles, knowing the velocity field, we integrate the set of equations

$$\frac{d\xi}{d\tau} = \mathbf{u}\left(\xi, \tau\right), \quad \xi\left(\tau\right)|_{\tau=t} = \mathbf{x}. \tag{3.19}$$

When the flow is steady, the velocity field is independent of time, and we have an *autonomous system* to integrate.

As an example, consider the case where the velocity is steady and homogeneous:

$$\mathbf{u}\left(\mathbf{x}\right) = \mathbf{Lx}. \tag{3.20}$$

The path lines are obtained by solving

$$\dot{\mathbf{x}} = \mathbf{Lx}, \quad \mathbf{x}\left(0\right) = \mathbf{X}. \tag{3.21}$$

The solution to this is

$$\mathbf{x} = \Phi\left(t\right)\mathbf{X}, \tag{3.22}$$

where Φ is called the *fundamental matrix*. It obeys

$$\frac{d\Phi}{dt} = \mathbf{L}\Phi, \quad \Phi\left(0\right) = \mathbf{I}. \tag{3.23}$$

In fact, we find from (3.13) that F is indeed the fundamental matrix of system (3.21). The solution to (3.23) is

$$\mathbf{F}\left(t\right) = \exp\left(t\mathbf{L}\right), \tag{3.24}$$

where the exponential function of a tensor is defined as [8]

$$\exp\left(\mathbf{A}\right) = \sum_{n=0}^{\infty} \frac{1}{n!}\mathbf{A}^n = \mathbf{I} + \mathbf{A} + \frac{1}{2}\mathbf{A}^2 + \cdots$$

Thus the path lines are all given by

$$\xi\left(\tau\right) = \Phi\left(\tau\right)\mathbf{X} = \Phi\left(\tau\right)\Phi\left(t\right)^{-1}\mathbf{x}\left(t\right),$$
$$= e^{\left(\tau-t\right)\mathbf{L}}\mathbf{x}. \tag{3.25}$$

The problem of calculating the exponential of a constant tensor falls into three categories, depending on the canonical form of the tensor, see Huilgol and Phan-Thien [7].

3.2.5 Oscillatory Shear Flow

We now consider the oscillatory shear flow, where the velocity depends explicitly on time:

$$u = \gamma_a y \cos \omega t, \quad v = 0, \quad w = 0. \tag{3.26}$$

The equations to solve for the path lines are

$$\dot{x} = \omega\gamma_a y \cos\omega t, \quad \dot{y} = 0, \quad \dot{z} = 0,$$
$$x(0) = X, \quad y(0) = Y, \quad z(0) = Z.$$

Integrate these,

$$x(t) = X + \gamma_a Y \sin\omega t, \quad y(t) = Y, \quad z(t) = Z.$$

The path lines are then given by, note that $\xi = (\xi, \psi, \zeta)$,

$$\xi(\tau) = x + \gamma_a y(\sin\omega\tau - \sin\omega t), \quad \psi(\tau) = y, \quad \zeta(\tau) = z. \tag{3.27}$$

Suppose the path lines have been determined, then the relative deformation gradient and the strain tensors may be calculated from (3.16) and (3.17). We illustrate this with the path lines for the simple shear flow (3.71):

$$\mathbf{F}_t(\tau) = (\nabla_\mathbf{x}\xi)^T = \mathbf{I} + (\tau - t)\mathbf{L},$$

which yields the Cauchy–Green strain tensor

$$\mathbf{C}_t(\tau) = \mathbf{F}_t(\tau)^T \mathbf{F}_t(\tau) = (\mathbf{I} + (\tau - t)\mathbf{L})^T (\mathbf{I} + (\tau - t)\mathbf{L})$$
$$= \mathbf{I} + (\tau - t)(\mathbf{L} + \mathbf{L}^T) + (\tau - t)^2 \mathbf{L}^T\mathbf{L}. \tag{3.28}$$

3.3 Rivlin–Ericksen Tensors

Suppose the relative Cauchy–Green tensor has been determined. The n-th *Rivlin–Ericksen tensor* is defined as

$$\mathbf{A}_n(t) = \left.\frac{d^n}{d\tau^n}\mathbf{C}_t(\tau)\right|_{\tau=t}, \quad n = 1, 2, \ldots \tag{3.29}$$

Since $\mathbf{C}_t(\tau)|_{\tau=t} = \mathbf{I}$, because the deformation gradient is the unit tensor at $\tau = t$, we may put

$$\mathbf{A}_0 = \mathbf{I}, \tag{3.30}$$

and extend the definition (3.29) to $n = 0, 1, \ldots$ In effect, the Rivlin–Ericksen tensors are defined as the coefficients in the Taylor series about t:

$$\mathbf{C}_t(\tau) = \sum_{n=0}^{\infty} \frac{(\tau - t)^n}{n!}\mathbf{A}_n(t). \tag{3.31}$$

Rivlin–Ericksen tensors can be determined directly from the velocity field, without having to find the strain tensor. This is shown below. First, we note from (3.18),

$$\frac{d}{d\tau}\mathbf{F}_t\left(\tau\right) = \frac{d}{d\tau}\left[\mathbf{F}\left(\tau\right)\mathbf{F}\left(t\right)^{-1}\right] = \mathbf{L}\left(\tau\right)\mathbf{F}\left(\tau\right)\mathbf{F}\left(t\right)^{-1} = \mathbf{L}\left(\tau\right)\mathbf{F}_t\left(\tau\right).$$

Thus

$$\frac{d}{d\tau}\mathbf{F}_t\left(\tau\right)^T = \left[\mathbf{L}\left(\tau\right)\mathbf{F}_t\left(\tau\right)\right]^T = \mathbf{F}_t\left(\tau\right)^T\mathbf{L}\left(\tau\right)^T.$$

Therefore, from (3.17)

$$\frac{d}{d\tau}\mathbf{C}_t\left(\tau\right) = \frac{d}{d\tau}\left[\mathbf{F}_t\left(\tau\right)^T\mathbf{F}_t\left(\tau\right)\right]$$
$$= \mathbf{F}_t\left(\tau\right)^T\mathbf{L}\left(\tau\right)^T\mathbf{F}_t\left(\tau\right) + \mathbf{F}_t\left(\tau\right)^T\mathbf{L}\left(\tau\right)\mathbf{F}_t\left(\tau\right). \tag{3.32}$$

As $\tau = t$, the relative deformation gradient is the unit tensor, and we have

$$\mathbf{A}_1\left(t\right) = \left.\frac{d}{d\tau}\mathbf{C}_t\left(\tau\right)\right|_{\tau=t} = \mathbf{L}\left(t\right) + \mathbf{L}\left(t\right)^T = 2\mathbf{D}\left(t\right), \tag{3.33}$$

i.e., the first Rivlin–Ericksen tensor is twice the strain rate tensor. Higher-order Rivlin–Ericksen tensors can be obtained by taking derivatives of (3.32) repeatedly. However, it is more instructive to look at an alternative way of calculating the Rivlin–Ericksen tensors, which also reveal the nature of the tensors.

We start with the length square of a fluid element in the current time:

$$dx\left(t\right)^2 = d\mathbf{x}\cdot d\mathbf{x} = \mathbf{F}^T\mathbf{F} : d\mathbf{X}d\mathbf{X} = \mathbf{C}\left(t\right) : d\mathbf{X}d\mathbf{X}.$$

Similarly

$$d\xi\left(\tau\right)^2 = \mathbf{C}\left(\tau\right) : d\mathbf{X}d\mathbf{X}.$$

Now, since

$$\mathbf{C}_t\left(\tau\right) = \mathbf{F}_t\left(\tau\right)^T\mathbf{F}_t\left(\tau\right) = \mathbf{F}\left(t\right)^{-T}\mathbf{F}\left(\tau\right)^T\mathbf{F}\left(\tau\right)\mathbf{F}\left(t\right)^{-1}$$
$$= \mathbf{F}\left(t\right)^{-T}\mathbf{C}\left(\tau\right)\mathbf{F}\left(t\right)^{-1},$$

we have

$$\frac{d^n}{d\tau^n}\mathbf{C}_t\left(\tau\right) = \mathbf{F}\left(t\right)^{-T}\frac{d^n}{d\tau^n}\mathbf{C}\left(\tau\right)\mathbf{F}\left(t\right)^{-1}.$$

Thus

$$\mathbf{F}\left(t\right)^T\left[\frac{d^n}{d\tau^n}\mathbf{C}_t\left(\tau\right)\right]\mathbf{F}\left(t\right) = \frac{d^n}{d\tau^n}\mathbf{C}\left(\tau\right).$$

Taking the scalar product of this with $d\mathbf{X}d\mathbf{X}$, we obtain

$$d\mathbf{X}^T \mathbf{F}(t)^T \left[\frac{d^n}{d\tau^n} \mathbf{C}_t(\tau) \right] \mathbf{F}(t) \, d\mathbf{X} = \left[\frac{d^n}{d\tau^n} \mathbf{C}(\tau) \right] : d\mathbf{X} d\mathbf{X}$$

$$\mathbf{A}_n : dxdx = \left. \frac{d^n}{d\tau^n} d\xi(\tau)^2 \right|_{\tau=t},$$

which relates the Rivlin–Ericksen tensors to the high-order stretching rate of a fluid element. For example,

$$\mathbf{A}_1 : dxdx = \frac{d}{dt} d\xi(t)^2 . \tag{3.34}$$

A recursive relation can be derived by noting that

$$\mathbf{A}_{n+1} : dxdx = \frac{d}{dt} \left(\frac{d^n}{dt^n} d\xi(t)^2 \right)$$

$$= \frac{d}{dt} (\mathbf{A}_n : dxdx)$$

$$= \left(\frac{d}{dt} \mathbf{A}_n : dxdx + \mathbf{A}_n : \frac{d}{dt}(dx) dx + \mathbf{A}_n : dx \frac{d}{dt}(dx) \right).$$

But

$$\frac{d}{dt}(dx) = \frac{d}{dt}(\mathbf{F} d\mathbf{X}) = \mathbf{L} \mathbf{F} d\mathbf{X} = \mathbf{L} dx,$$

and therefore (if uncomfortable with direct notation, do it in subscript notation)

$$\mathbf{A}_{n+1} : dxdx = \left(\frac{d}{dt} \mathbf{A}_n + \mathbf{A}_n \mathbf{L} + \mathbf{L}^T \mathbf{A}_n \right) : dxdx$$

which leads to the recursive formula due to Rivlin and Ericksen [9]

$$\mathbf{A}_{n+1} = \frac{d}{dt} \mathbf{A}_n + \mathbf{A}_n \mathbf{L} + \mathbf{L}^T \mathbf{A}_n, \quad \mathbf{A}_0 = \mathbf{I}, \quad n = 1, 2, \ldots \tag{3.35}$$

As an example, let's calculate the Rivlin–Ericksen tensors for the simple shear flow (3.69). The first Rivlin–Ericksen tensor is twice the strain rate tensor:

$$[\mathbf{A}_1] = \begin{bmatrix} 0 & \dot\gamma & 0 \\ \dot\gamma & 0 & 0 \\ 0 & 0 & 0 \end{bmatrix} .$$

The second Rivlin–Ericksen tensor is obtained from the first using (3.35),

$$\mathbf{A}_2 = \mathbf{A}_1 \mathbf{L} + \mathbf{L}^T \mathbf{A}_1$$

$$= \begin{bmatrix} 0 & \dot\gamma & 0 \\ \dot\gamma & 0 & 0 \\ 0 & 0 & 0 \end{bmatrix} \begin{bmatrix} 0 & \dot\gamma & 0 \\ 0 & 0 & 0 \\ 0 & 0 & 0 \end{bmatrix} + \begin{bmatrix} 0 & 0 & 0 \\ \dot\gamma & 0 & 0 \\ 0 & 0 & 0 \end{bmatrix} \begin{bmatrix} 0 & \dot\gamma & 0 \\ \dot\gamma & 0 & 0 \\ 0 & 0 & 0 \end{bmatrix} = \begin{bmatrix} 0 & 0 & 0 \\ 0 & 2\dot\gamma^2 & 0 \\ 0 & 0 & 0 \end{bmatrix} .$$

All other higher-order Rivlin–Ericksen tensors are zero for this flow. In fact, in this flow $\mathbf{L}^2 = \mathbf{0}$, and it can be verified from (3.28) that

$$\mathbf{C}_t\left(\tau\right) = \mathbf{I} + \left(\tau - t\right)\mathbf{A}_1 + \frac{1}{2}\left(\tau - t\right)^2 \mathbf{A}_2. \tag{3.36}$$

3.4 Small Strain

When the strain is small, in the sense that the fluid particles remain close to their original positions in the reference configuration at all times, then the strain may be calculated by introducing the displacement function $\mathbf{v}$:

$$\mathbf{v} = \mathbf{x}\left(\mathbf{X}, t\right) - \mathbf{X}. \tag{3.37}$$

The deformation gradient is

$$\mathbf{F} = \left(\nabla_{\mathbf{X}}\mathbf{x}\right)^T = \mathbf{I} + \mathbf{E}, \quad \mathbf{E} = \left(\nabla_{\mathbf{X}}\mathbf{v}\right)^T. \tag{3.38}$$

When the displacement gradient is small, terms of order $\varepsilon^2 = O\left(\|\mathbf{E}\|^2\right)$ and higher can be neglected, one has

$$\begin{aligned}
\mathbf{F}^{-1} &= \mathbf{I} - \mathbf{E}, \\
\mathbf{F}_t\left(\tau\right) &= \mathbf{F}\left(\tau\right)\mathbf{F}\left(t\right)^{-1} = \mathbf{I} + \mathbf{E}\left(\tau\right) - \mathbf{E}\left(t\right), \\
\mathbf{C}_t\left(\tau\right) &= \mathbf{F}_t\left(\tau\right)^T \mathbf{F}_t\left(\tau\right) = \mathbf{I} + \mathbf{E}\left(\tau\right) + \mathbf{E}\left(\tau\right)^T - \mathbf{E}\left(t\right) - \mathbf{E}\left(t\right)^T.
\end{aligned} \tag{3.39}$$

In terms of the infinitesimal strain tensor

$$\varepsilon = \frac{1}{2}\left(\mathbf{E} + \mathbf{E}^T\right), \tag{3.40}$$

we have

$$\mathbf{C} = \mathbf{I} + 2\varepsilon, \quad \mathbf{C}_t\left(\tau\right) = \mathbf{I} + 2\varepsilon\left(\tau\right) - 2\varepsilon\left(t\right). \tag{3.41}$$

In the polar decomposition of $\mathbf{F}$, $\mathbf{F} = \mathbf{R}\mathbf{U}$, $\mathbf{U}$ is the square root of $\mathbf{C}$,

$$\mathbf{U} = \mathbf{C}^{1/2} = \mathbf{I} + \varepsilon. \tag{3.42}$$

Thus

$$\mathbf{R} = \mathbf{F}\mathbf{U}^{-1} = \left(\mathbf{I} + \mathbf{E}\right)\left(\mathbf{I} - \varepsilon\right) = \mathbf{I} + \mathbf{E} - \varepsilon = \mathbf{I} + \omega,$$

where ω is the infinitesimal rotation tensor:

$$\omega = \frac{1}{2}\left(\mathbf{E} - \mathbf{E}^T\right). \tag{3.43}$$

3.5 Equations of Balance

The equations of balance are mathematical statements of the conservation of mass, linear and angular momentum, and energy.

3.5.1 Reynolds Transport Theorem

Theorem 1. Let $\Phi(\mathbf{x}, t)$ be a field (scalar, vector or tensor) defined over a region V occupied by the body $\mathcal{B}$ at time t. The Reynolds[1] transport theorem states that

$$\frac{d}{dt} \int_V \Phi dV = \int_V \left(\frac{d\Phi}{dt} + \Phi \nabla \cdot \mathbf{u} \right) dV = \int_V \left(\frac{\partial \Phi}{\partial t} + \Phi \nabla \cdot (\Phi \mathbf{u}) \right) dV, \tag{3.44}$$

where $\mathbf{u}$ is the velocity field, and d/dt is the material derivative (3.7). This is proved by expressing the volume integral in the reference configuration,

$$\frac{d}{dt} \int_V \Phi d\mathbf{x} = \frac{d}{dt} \int_{V_0} \Phi J d\mathbf{X}, \tag{3.45}$$

where $J = \det(\partial \mathbf{x}/\partial \mathbf{X}) = \det \mathbf{F}$ is the Jacobian of the transformation (we use $d\mathbf{x}$ interchangeably with dV), and V_0 is the region occupied by the reference configuration.

Lemma. We record the following Lemma

$$\frac{dJ}{dt} = J \nabla \cdot \mathbf{u}. \tag{3.46}$$

This is proved by using a result obtained previously, §2.8.1, or by noting that

$$\begin{aligned} dJ &= \det(\mathbf{F} + d\mathbf{F}) - \det(\mathbf{F}) = \det\left[\mathbf{F}\left(\mathbf{I} + \mathbf{F}^{-1} d\mathbf{F} \right) \right] - \det \mathbf{F} \\ &= \det \mathbf{F} \det\left(\mathbf{I} + \mathbf{F}^{-1} d\mathbf{F} \right) - \det \mathbf{F} \\ &= \det \mathbf{F}\left(1 + \mathrm{tr}\left(\mathbf{F}^{-1} d\mathbf{F} \right) \right) - \det \mathbf{F} \\ &= J \mathrm{tr}\left(\mathbf{F}^{-1} d\mathbf{F} \right). \end{aligned}$$

Divide both sides by dt, and using $\dot{\mathbf{F}} = \mathbf{L}\mathbf{F}$, and

$$\mathrm{tr}\left(\mathbf{F}^{-1} \mathbf{L} \mathbf{F} \right) = F_{ij}^{-1} L_{jk} F_{ki} = \delta_{kj} L_{jk} = L_{kk} = \nabla \cdot \mathbf{u}.$$

This lemma can be used directly in (3.45) to prove (3.44).

---

[1] Osborne Reynolds (1842–1912) introduced the lubrication theory and formulated the framework for turbulence flow. The Reynolds number and stresses are named after him.

Another form for the Reynolds transport theorem which emphasizes the flux of Φ into the volume V bounded by the surface S is given below by recognizing that

$$\int_V \left(\frac{d\Phi}{dt} + \Phi\nabla \cdot \mathbf{u}\right) dV = \int_V \left(\frac{\partial\Phi}{\partial t} + \mathbf{u} \cdot \nabla\Phi + \Phi\nabla \cdot \mathbf{u}\right) dV$$

$$= \int_V \left(\frac{\partial\Phi}{\partial t} + \nabla \cdot (\Phi\mathbf{u})\right) dV$$

$$= \int_V \frac{\partial\Phi}{\partial t} dV + \int_S (\Phi\mathbf{u}) \cdot \mathbf{n} dS$$

Theorem 2. Thus

$$\frac{d}{dt}\int_V \Phi d\mathbf{x} = \int_V \frac{\partial\Phi}{\partial t} dV + \int_S (\Phi\mathbf{u}) \cdot \mathbf{n} dS. \tag{3.47}$$

This form allows a physical interpretation of the theorem: the first term on the right represents the rate of creation of the quantity Φ, and the second term, the flux of Φ into the volume V through its bounding surface.

3.5.2 Conservation of Mass

The mass in the volume V is conserved at all time, i.e.,

$$\frac{d}{dt}\int_V \rho dV = 0,$$

where $\rho(\mathbf{x}, t)$ is the density field at time t. From Reynolds transport theorem (3.44),

$$\int_V \left(\frac{\partial\rho}{\partial t} + \nabla \cdot (\rho\mathbf{u})\right) dV = 0.$$

Since the volume V is arbitrary, a necessary and sufficient condition for the conservation of mass is

$$\frac{\partial\rho}{\partial t} + \nabla \cdot (\rho\mathbf{u}) = \frac{d\rho}{dt} + \rho\nabla \cdot \mathbf{u} = 0. \tag{3.48}$$

For an incompressible material, the density is constant everywhere, and the conservation of mass demands that

$$\nabla \cdot \mathbf{u} = \frac{\partial u_i}{\partial x_i} = \mathrm{tr}\mathbf{L} = \mathrm{tr}\mathbf{D} = \frac{1}{2}\mathrm{tr}\mathbf{A}_1 = 0. \tag{3.49}$$

From a solid point of view, the conservation of mass requires that

$$\rho dV = \rho_R dV_R \quad \therefore \quad \rho_R = \rho J,$$

where $J = \det \mathbf{F}$ is the Jacobian of the deformation. For an incompressible material, we shall demand that

$$J = \det \mathbf{F} = 1, \quad \det \mathbf{C} = 1, \tag{3.50}$$

at all time.

Theorem 3. As a corollary to (3.44) and (3.48) we have

$$\frac{d}{dt}\int_V \rho\Phi dV = \int_V \rho\frac{d\Phi}{dt}dV. \tag{3.51}$$

This is easily demonstrated by using (3.44) and (3.48) on the left hand side:

$$\frac{d}{dt}\int_V \rho\Phi dV = \int_V \left(\Phi\left(\frac{d\rho}{dt} + \rho\nabla\cdot\mathbf{u}\right) + \rho\frac{d\Phi}{dt}\right)dV.$$

3.5.3 Conservation of Momentum

The forces acting on the body are either surface forces $\mathbf{t}$ (or tractions), and body forces $\mathbf{b}$ (those that act at a distance).

Body Force Density. An example of body force is gravitational. If $\mathbf{b}(\mathbf{x}, t)$ is the body force density defined on V, then the resulting force and moment (about O) on V due to the body force field are given respectively by

$$\int_V \rho\mathbf{b}dV, \qquad \int_V \mathbf{x}\times\mathbf{b}dV.$$

Surface Force. Surface traction is a concept due to Cauchy. Consider a particle X occupying the position $\mathbf{x}$ at time t. Construct a surface S_t through this point with unit normal vector $\mathbf{n}(\mathbf{x}, t)$ at point $\mathbf{x}$, which separates the body into two regions: B^+ is the region into which the unit normal $\mathbf{n}$ is directed and B^- on the other side (see Fig. 3.5). $\mathbf{t}(\mathbf{x}, t; \mathbf{n})$ is called the surface force density per unit (current) area if the force and moment (about O) exerted on B^- by B^+ are given respectively by

$$\int_{S_t} \mathbf{t}(\mathbf{x}, t)\, dS, \qquad \int_{S_t} \mathbf{x}\times\mathbf{t}(\mathbf{x}, t)\, dS.$$

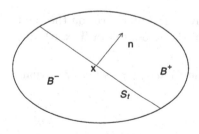

Fig. 3.5. The traction $\mathbf{t}(\mathbf{x}, t; \mathbf{n})$ is the force per unit area exerted by B^+ on B^-

Balance of Linear Momentum. Newton's second law (a postulate), as applied to a volume V occupied by the body, requires that

$$\frac{d}{dt}\int_V \rho \mathbf{u} dV = \int_S \mathbf{t} dS + \int_V \rho \mathbf{b} dV. \tag{3.52}$$

Here S is the bounding surface of V. The first term on the right hand of (3.52) represents the surface force acting on V due the body outside V, and the second term is the net body force on V.

Corollary (3.51) can be used on the left side of the preceding equation, leading to

$$\int_V \rho \frac{d\mathbf{u}}{dt} dV = \int_S \mathbf{t} dS + \int_V \rho \mathbf{b} dV. \tag{3.53}$$

Balance of Angular Momentum. The balance of angular momentum (a postulate) can be likewise written down as, in the absence of body couple,

$$\frac{d}{dt}\int_V \mathbf{x} \times \rho \mathbf{u} dV = \int_S \mathbf{x} \times \mathbf{t} dS + \int_V \mathbf{x} \times \rho \mathbf{b} dV. \tag{3.54}$$

Again, corollary (3.51) can be used on the left side of the preceding equation, leading to

$$\int_V \mathbf{x} \times \rho \frac{d\mathbf{u}}{dt} dV = \int_S \mathbf{x} \times \mathbf{t} dS + \int_V \mathbf{x} \times \rho \mathbf{b} dV. \tag{3.55}$$

Note that the term $d\mathbf{x}/dt = \mathbf{u}$ does not contribute to this because of the cross product term. This should be contrasted to the rigid body mechanics case where the Newton's second law is a postulate and the balance of angular momentum is a consequence (a theorem). Here two postulates are required.

The term

$$\mathbf{a} = \frac{d\mathbf{u}}{dt} = \frac{\partial \mathbf{u}}{\partial t} + \mathbf{u} \cdot \nabla \mathbf{u} \tag{3.56}$$

is recognised as the acceleration field. To convert the first term on the right into a volume integral, we need the concept of the stress tensor $\mathbf{T}(\mathbf{x}; \mathbf{n})$, due to Cauchy.

Cauchy Stress Tensor. The existence of the *Cauchy stress tensor* is guaranteed by the following theorem.

• The traction vector satisfies

$$\mathbf{t}(\mathbf{x}, t; -\mathbf{n}) = -\mathbf{t}(\mathbf{x}, t; \mathbf{n}). \tag{3.57}$$

• Further, there exists a second-order tensor field $\mathbf{T}(\mathbf{x}, t)$ with the following properties:

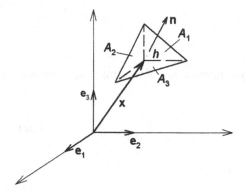

Fig. 3.6. Existence of the stress tensor

$$t\left(\mathbf{x}, t; \mathbf{n}\right) = \mathbf{T}\left(\mathbf{x}, t\right) \cdot \mathbf{n}, \tag{3.58}$$

with components in the frame $\mathcal{F} = \{\mathbf{e}_1, \mathbf{e}_2, \mathbf{e}_3\}$

$$T_{ij}\left(\mathbf{x}, t\right) = t\left(\mathbf{x}, t; \mathbf{e}_j\right) \cdot \mathbf{e}_i. \tag{3.59}$$

The proof, due to Gurtin [1] lies in the construction of a one-parameter familly of tetrahedra, Fig. 3.6, with vertex at point $\mathbf{x}$, height h (the parameter). The face normal to $\mathbf{n}$ has area A; the face normal to $-\mathbf{e}_i$ has area A_i. From the directional cosine of $\mathbf{n}$

$$A_i(h) = A(h)n_i.$$

Furthermore the volume of the tetrahedron is $V_h = \frac{1}{3}hA(h)$.

Applying (3.53) to the tetrahedron (omitting the time argument t for brevity):

$$\int_{V_h} \rho a\left(\mathbf{y}\right) dV\left(\mathbf{y}\right) = \int_{V_h} \rho b\left(\mathbf{y}\right) dV\left(\mathbf{y}\right) + \int_{A(h)} t\left(\mathbf{y}; \mathbf{n}\right) dS\left(\mathbf{y}\right)$$
$$+ \sum_{j=1}^{3} \int_{A_i(h)} t\left(\mathbf{y}; -\mathbf{e}_j\right) dS\left(\mathbf{y}\right).$$

From the continuity of all the field variables, and the mean-value theorem

$$\left(\rho a\left(\mathbf{x}\right) + \mu\right) V_h = \left(\rho b\left(\mathbf{x}\right) + \alpha\right) V_h + \left(t\left(\mathbf{x}; \mathbf{n}\right) + \beta\right) A\left(h\right)$$
$$+ \sum_{j=1}^{3} \left[t\left(\mathbf{x}; -\mathbf{e}_j\right) + \beta_j\right] A\left(h\right) n_j, \tag{3.60}$$

where $\alpha\left(h\right), \beta\left(h\right), \beta_j\left(h\right), \mu\left(h\right) = o\left(1\right),\ h \to 0$. Divide (3.60) by $A\left(h\right)$ and let $h \to 0$, we find

$$t\left(\mathbf{x}; \mathbf{n}\right) = -\sum_{j=1}^{3} t\left(\mathbf{x}; -\mathbf{e}_j\right) n_j. \tag{3.61}$$

In the case where $\mathbf{n} = \mathbf{e}_i$ (fixed i)

$$\mathbf{t}(\mathbf{x}; -\mathbf{e}_i) = -\mathbf{t}(\mathbf{x}; \mathbf{e}_i),$$

which leads to (3.57) due the arbitrary configuration of the frame of reference. Thus

$$\mathbf{t}(\mathbf{x}; \mathbf{n}) = \sum_{j=1}^{3} \mathbf{t}(\mathbf{x}; \mathbf{e}_j) n_j. \tag{3.62}$$

From (3.60), the components of $\mathbf{t}$ are given by

$$t_i(\mathbf{x}; \mathbf{n}) = \mathbf{t}(\mathbf{x}; \mathbf{n}) \cdot \mathbf{e}_i = \sum_{j=1}^{3} \mathbf{t}(\mathbf{x}; \mathbf{e}_j) \cdot \mathbf{e}_i n_j$$

$$= T_{ij}(\mathbf{x}) n_j, \tag{3.63}$$

where T_{ij} are defined as in (3.59). From the quotient rule (§2.3.10), T_{ij} are indeed the components of a second-order tensor thus proving the existence of the stress tensor (3.58).

Conservation of Linear Momentum. Returning to the balance of linear momentum (3.53) and using the definition of the stress tensor (3.58),

$$\int_V \rho \frac{d\mathbf{u}}{dt} dV = \int_S \mathbf{T} \cdot \mathbf{n} dS + \int_V \rho \mathbf{b} dV,$$

$$\int_V \rho \frac{du_i}{dt} dV = \int_S T_{ij} n_j dS + \int_V \rho b_i dV.$$

The surface integral on the right hand of the preceding equation can be converted into a volume integral to obtain

$$\int_V \rho \frac{d\mathbf{u}}{dt} dV = \int_V \nabla \mathbf{T}^T dV + \int_V \rho \mathbf{b} dV,$$

$$\int_V \rho \frac{du_i}{dt} dV = \int_S \frac{\partial T_{ij}}{\partial x_j} dS + \int_V \rho b_i dV.$$

Since the integrand is continuous on an arbitrary V, the conservation of linear momentum becomes

$$\rho \frac{d\mathbf{u}}{dt} = \nabla \mathbf{T}^T + \rho \mathbf{b}, \qquad \rho \frac{du_i}{dt} = \frac{\partial T_{ij}}{\partial x_j} + \rho b_i. \tag{3.64}$$

Conservation of Angular Momentum. Returning to the balance of angular momentum, 3.54,

$$\int_V \mathbf{x} \times \rho \frac{d\mathbf{u}}{dt} dV = \int_S \mathbf{x} \times \mathbf{t} dS + \int_V \mathbf{x} \times \rho \mathbf{b} dV$$

and examine the first term on the right,

$$\int_S \varepsilon_{ijk} x_j t_k dS = \int_S \varepsilon_{ijk} x_j T_{kl} n_l dS = \int_V \frac{\partial}{\partial x_l} \left(\varepsilon_{ijk} x_j T_{kl} \right) dV$$

$$= \int_V \left(\varepsilon_{ijk} \delta_{jl} T_{kl} + \varepsilon_{ijk} x_j \frac{\partial T_{kl}}{\partial x_l} \right) dV$$

$$= \int_V \left(t_i^A + \varepsilon_{ijk} x_j \frac{\partial T_{kl}}{\partial x_l} \right) dV,$$

where the "axial vector" is defined as $t_i^A = \varepsilon_{ijk} T_{kj}$. When these results are substituted back into (3.54),

$$\int_V \mathbf{x} \times \rho \frac{d\mathbf{u}}{dt} dV = \int_V \left(\mathbf{t}^A + \mathbf{x} \times \nabla \cdot \mathbf{T}^T \right) dV + \int_V \mathbf{x} \times \rho \mathbf{b} dV$$

$$\int_V \mathbf{t}^A dV = \int_V \mathbf{x} \times \left[\rho \frac{d\mathbf{u}}{dt} - \nabla \cdot \mathbf{T}^T - \rho \mathbf{b} \right] dV.$$

From the conservation of linear momentum, and the continuity of the integrands in the arbitrary volume V, it follows that

$$\mathbf{t}^A = 0 \quad \therefore \quad \varepsilon_{ijk} T_{kj} = 0 \quad \therefore \quad \mathbf{T} = \mathbf{T}^T \text{ or } \mathbf{T} \text{ is symmetric.} \qquad (3.65)$$

A necessary and sufficient condition for the balance of angular momentum, in the absence of body couples, is that the stress be symmetric. In deriving (3.67), both the conservation of mass and linear momentum are needed.

3.5.4 Conservation of Energy

We define the kinetic energy K, and internal energy E as

$$K = \int_V \frac{1}{2} \rho u^2 dV, \quad E = \int_V \rho \varepsilon dV, \qquad (3.66)$$

where ε is the specific internal energy per unit mass. The rate of work done on the body due to surface and body forces is given by

$$\int_S \mathbf{t} \cdot \mathbf{u} dS + \int_V \rho \mathbf{b} \cdot \mathbf{u} dV.$$

If we define $\mathbf{q}$ to be the flux of energy out of S per unit area, and r the amount of energy created per unit mass, then the mathematical statement for the first law of thermodynamics can be expressed as

$$\frac{d}{dt} \int_V \left(\frac{1}{2} \rho u^2 + \rho \varepsilon \right) dV = \int_S \left(\mathbf{t} \cdot \mathbf{u} - \mathbf{q} \cdot \mathbf{n} \right) dS$$

$$+ \int_V \left(\rho r + \rho \mathbf{b} \cdot \mathbf{u} \right) dV. \qquad (3.67)$$

The left of (3.67) can be expressed as, using Reynolds transport theorem 3,

$$\frac{d}{dt}\int_V \left(\frac{1}{2}\rho u^2 + \rho\varepsilon\right)dV = \int_V \rho\left(\mathbf{u}\cdot\mathbf{a} + \dot\varepsilon\right)dV.$$

The surface integral on the right of (3.67) is converted into volume integral as

$$\int_S (\mathbf{t}\cdot\mathbf{u} - \mathbf{q}\cdot\mathbf{n})\,dS = \int_V \left(\frac{\partial}{\partial x_j}(u_i T_{ij}) - \frac{\partial q_i}{\partial x_i}\right)dV$$

$$= \int_V (\mathbf{u}\cdot\nabla\mathbf{T}^T + \mathbf{T}:\mathbf{L} - \nabla\cdot\mathbf{q})\,dV.$$

Hence (3.67) becomes

$$\int_V \rho\left(\mathbf{u}\cdot\mathbf{a} + \dot\varepsilon\right)dV = \int_V (\mathbf{u}\cdot\nabla\mathbf{T}^T + \mathbf{T}:\mathbf{L} - \nabla\cdot\mathbf{q})\,dV$$

$$+ \int_V (\rho r + \rho\mathbf{b}\cdot\mathbf{u})\,dV.$$

Because of the conservation of linear momentum, and the contunuity of the integrands in the arbitrary volume V, the conservation of energy is reduced to

$$\rho\dot\varepsilon = \mathbf{T}:\mathbf{D} - \nabla\cdot\mathbf{q} + \rho r. \tag{3.68}$$

In deriving this, all the three balance equations for mass, linear momentum and angular momentum are required. The term $\mathbf{T}:\mathbf{D}$ represents the rate of work done by the stress, or the "stress power". It is seen that

$$\mathbf{T}:\mathbf{D} = -P\operatorname{tr}\mathbf{D} + \mathbf{S}:\mathbf{D}.$$

The rate of work done by the pressure for an incompressible fluid is zero, because $\operatorname{tr}\mathbf{D} = 0$.

Problems

Problem 3.1 Using $\mathbf{F}\mathbf{F}^{-1} = \mathbf{I}$, show that

$$\frac{d}{dt}\mathbf{F}^{-1} = -\mathbf{F}\mathbf{L}, \quad \mathbf{F}^{-1}(0) = \mathbf{I}.$$

Problem 3.2 For a simple shear flow, where the velocity field takes the form

$$u = \dot\gamma y, \quad v = 0, \quad w = 0, \tag{3.69}$$

show that the velocity gradient and its exponent are given by

$$[\mathbf{L}] = \begin{bmatrix} 0 & \dot{\gamma} & 0 \\ 0 & 0 & 0 \\ 0 & 0 & 0 \end{bmatrix}, \quad \exp(\mathbf{L}) = \mathbf{I} + \mathbf{L}. \tag{3.70}$$

Show that the path lines are given by

$$\xi(\tau) = \mathbf{x} + (\tau - t)\,\mathbf{Lx}. \tag{3.71}$$

so that a fluid element dX can only be stretched linearly in time at most.

Problem 3.3 Repeat the same exercise for an elongational flow, where

$$u = ax, \quad v = by, \quad w = cz, \quad a + b + c = 0. \tag{3.72}$$

In this case, show that

$$[\mathbf{L}] = \begin{bmatrix} a & 0 & 0 \\ 0 & b & 0 \\ 0 & 0 & c \end{bmatrix}, \quad [e^{\mathbf{L}}] = \begin{bmatrix} e^a & 0 & 0 \\ 0 & e^b & 0 \\ 0 & 0 & e^c \end{bmatrix}. \tag{3.73}$$

Show that the path lines are given by

$$[\xi(\tau)] = \begin{bmatrix} \xi \\ \psi \\ \zeta \end{bmatrix} = \begin{bmatrix} e^{a(\tau-t)} & 0 & 0 \\ 0 & e^{b(\tau-t)} & 0 \\ 0 & 0 & e^{c(\tau-t)} \end{bmatrix} \begin{bmatrix} x \\ y \\ z \end{bmatrix}. \tag{3.74}$$

Conclude that exponential flow can stretch the fluid element exponentially fast.

Problem 3.4 Consider a super-imposed oscillatory shear flow:

$$u = \dot{\gamma}_m y, \quad v = 0, \quad w = \omega\gamma_a y \cos\omega t.$$

Show that the path lines are

$$\begin{aligned} \xi(\tau) &= x + \dot{\gamma}_m\,(\tau - t)\,y, \\ \psi(\tau) &= y, \\ \zeta(\tau) &= z + \gamma_0 y\,(\sin\omega\tau - \sin\omega t). \end{aligned} \tag{3.75}$$

Problem 3.5 Calculate the Rivlin–Ericksen tensors for the elongational flow (3.72).

Problem 3.6 Calculate the Rivlin–Ericksen tensors for the unsteady flow (3.75).

Problem 3.7 Write down, in component forms the conservation of mass and linear momentum equations, assuming the fluid is incompressible, in Cartesian, cylindrical and spherical coordinate systems.

4. Constitutive Equation: General Principles

Basic principles and some classical constitutive equations

In isothermal flow where the conservation of energy is not relevant, there are four balance equations (one conservation of mass and three conservation of linear momentum), and there are 10 scalar variables (3 velocity, one pressure, and 6 independent stress components – thanks to the conservation of angular momentum). Clearly we do not have a mathematically well-posed problem until 6 extra equations are found. The constitutive equation, or the rheological equation of state, provides the linkage between the stresses and the kinematics and provides the missing information. Modelling a complex fluid, or finding a relevant constitutive equation for the fluid, is the central concern in rheology. In this chapter, we review some of the well-known classical models, and the general principles underlying constitutive modelling.

4.1 Some Well-Known Constitutive Equations

4.1.1 Perfect Gas

The most well known constitutive equation is the perfect gas law, due to Boyle (Fig. 4.1), where the state of the gas is fully specified by its volume V, its pressure P and its temperature T

$$PV = RT, \tag{4.1}$$

where R is a universal gas constant.

Fig. 4.1. Robert Boyle (1627–1691) made several important contributions to Physics and Chemistry, the best known is the Perfect Gas Law. He employed Robert Hooke as his assistant in the investigation of the behaviour of air. His experiments led him to believe in vacuum, and reject Descartes'concept of ether

4.1.2 Inviscid Fluid

The perfect fluid concept of D'Alembert (Fig. 4.2) and Euler (Fig. 3.2) is another well-known constitutive equation. In our notation, the stress is given by

$$\mathbf{T} = -P\mathbf{I}, \quad T_{ij} = -P\delta_{ij}. \tag{4.2}$$

Here, P is the pressure. The inviscid fluid model fails to account for the pressure losses in pipe flow, and a better model is needed.

Fig. 4.2. Jean d'Alembert (1717–1783) was a French mathematician. He made several important contributions to Mechanics. He is most famous for the D'Alembert Principle

4.1.3 Fourier's Law

Students of thermodynamics would recognize Fourier's law of heat conduction, linking the heat transfer rate $\mathbf{q}$ to the temperature gradient:

$$\mathbf{q} = -k\nabla\theta, \quad q_i = -k\frac{\partial\theta}{\partial x_i}, \tag{4.3}$$

where k is the thermal conductivity and θ is the temperature field.

Fig. 4.3. Joseph Fourier (1768–1830) was a French historian, administrator and mathematician. He was famous for his Fourier's series. His name is one of the 72 inscribed on the Eiffel tower

4.1.4 Hookean Solid

In 1678 Robert Hooke published his now famous law for material behaviour as a solution to an anagram that he published two years earlier *ut tensio sic vis*, which roughly translated as extension is proportional to the force (see Fig. 4.4). This idea has gone through several revisions by several well-known

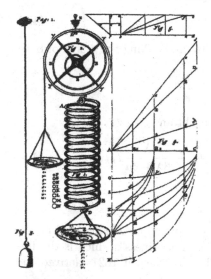

Fig. 4.4. Robert Hooke (1635–1703) was perhaps the foremost experimental scientist in the 17th century. He was a noted architect, an inventor (Hooke universal joint, spring control in watches, reflecting telescope, etc.), a mathematician, a physicist, a chemist and contributed significantly to anatomy, astronomy, botany, chemistry – the term "cell" is due to him. No portrait survived him. He is most well-known for the law of elasticity that bears his name

Fig. 4.5. C.L.M.H Navier (1785–1836), a French engineer, developed a particle model for an elastic solid, which has a shear modulus and a Poisson's ratio of 0.25 (see Love 1944). He obtained the Navier–Stokes equations by molecular arguments. His name is one of the 72 inscribed on the Eiffel tower

scientists, including Young, Poisson, and Navier (Fig. 4.5) who thought that one needs only one elastic constant. The concept of the stress tensor was introduced by Cauchy (Fig. 3.4), who also gave the correct version of the constitutive equation for infinitesimal elasticity in 1827. In our notation, the stress tensor is given by

$$\mathbf{T} = \mathbf{C} \cdot \varepsilon, \quad T_{ij} = C_{ijkl}\varepsilon_{kl}, \tag{4.4}$$

where

$$\varepsilon = \frac{1}{2}\left(\nabla\mathbf{v} + \nabla\mathbf{v}^T\right), \quad \varepsilon_{ij} = \frac{1}{2}\left(\frac{\partial v_i}{\partial x_j} + \frac{\partial v_j}{\partial x_i}\right) \tag{4.5}$$

is the infinitesimal strain tensor, $\mathbf{v} = \mathbf{x} - \mathbf{X}$ is the displacement vector, $\mathbf{C}$ is a 4th order elasticity tensor, a material tensorial constant, which reflects all the anisotropy of the material. For an isotropic solid, the elastic constants are reduced to just two:

$$C_{ijkl} = \lambda\delta_{ij}\delta_{kl} + \mu\left(\delta_{ik}\delta_{jl} + \delta_{il}\delta_{jk}\right), \tag{4.6}$$

where λ and μ are called Lamé moduli. Using this in (4.4) produces

$$\mathbf{T} = \lambda\,(\mathrm{tr}\,\varepsilon)\,\mathbf{I} + 2\mu\varepsilon, \quad T_{ij} = \lambda\varepsilon_{kk}\delta_{ij} + 2\mu\varepsilon_{ij}. \tag{4.7}$$

An alternative form for this can be derived by first taking the traces of both sides of (4.7)

$$\mathrm{tr}\,\mathbf{T} = (3\lambda + 2\mu)\,\mathrm{tr}\,\varepsilon,$$

and then substituting this back into (4.7) to obtain

$$\varepsilon = \frac{1}{2\mu}\mathbf{T} - \frac{\lambda}{3\lambda + 2\mu}\,(\mathrm{tr}\,\mathbf{T})\,\mathbf{I}, \quad \varepsilon_{ij} = \frac{1}{2\mu}T_{ij} - \frac{\lambda}{3\lambda + 2\mu}T_{kk}\delta_{ij}. \tag{4.8}$$

4.1.5 Newtonian Fluid

About nine years after Hooke published his paper on elasticity, Newton (Fig. 4.6) introduced the concept of "lack of slipperiness", which is the important quantity that we call viscosity. Then Navier, in 1827, derived the Navier–Stokes equation,

$$\rho\frac{d\mathbf{u}}{dt} = -\nabla P + \eta\nabla^2\mathbf{u} + \rho\mathbf{b}. \tag{4.9}$$

In this equation, the terms $\eta\nabla^2\mathbf{u}$ represent the viscous forces, although Navier did not attach much physical significance to η. Stokes (Fig. 4.7) gave the correct form to the constitutive equation that we call a Newtonian fluid:

$$\mathbf{T} = -P\mathbf{I} + \Lambda\,(\mathrm{tr}\,\mathbf{D})\,\mathbf{I} + 2\eta\mathbf{D}, \quad T_{ij} = -P\delta_{ij} + \Lambda D_{kk}\delta_{ij} + 2\eta D_{ij}. \tag{4.10}$$

Here, η is the viscosity, Λ is the bulk viscosity, and $\mathbf{D} = (\nabla\mathbf{u} + \nabla\mathbf{u}^T)/2$ is the strain rate tensor. Stokes assumed that

$$\Lambda = -\frac{2}{3}\eta,$$

Fig. 4.6. Sir Isaac Newton (1643–1727) was a dominating personality in Science. He (and concurrently Leibnitz) invented differential and integral calculus, and the gravitational theory. He was appointed the Lucasian Professor at Cambridge at the age of 26

Fig. 4.7. George Gabrielle Stokes (1819–1903) was a Irish mathematician. He was appointed the Lucasian Professor at Cambridge at the age of 30. He is remembered for Stokes flow and his contributions in the Navier–Stokes equations

so that pure volumetric change does not affect the stress ($\operatorname{tr}\mathbf{T}$ is independent of $\operatorname{tr}\mathbf{D}$). Furthermore, the terms $\Lambda\operatorname{tr}\mathbf{D}\delta_{ij}$ can be absorbed in the pressure term. This leads to the familiar constitutive equation for Newtonian fluids:

$$\mathbf{T} = -P\mathbf{I} + 2\eta\mathbf{D}. \tag{4.11}$$

4.1.6 Non-Newtonian Fluid

The term *non-Newtonian fluid* is an all-encompassing term denoting any fluid that does not obey (4.11). To discuss constitutive relations for non-Newtonian fluids, we need a convenient way to classify different flow regimes.

4.2 Weissenberg and Deborah Numbers

Most non-Newtonian fluids have a characteristic time scale λ. In a flow with a characteristic shear rate $\dot{\gamma}$ and a characteristic frequency ω, or characteristic time T, two dimensionless groups can be formed

$$\begin{aligned}
&\text{Deborah number} &&\text{De} = \lambda\omega \text{ or } \lambda/T, \\
&\text{Weissenberg number} &&\text{Wi} = \lambda\dot{\gamma}
\end{aligned} \tag{4.12}$$

4.2.1 Deborah Number

The Deborah number,[1] the ratio between the fluid relaxation time and the flow characteristic time, represents the transient nature of the flow relative to the fluid time scale. If the observation time scale is large (small De number), the material responses like a fluid, and if it is small (large De number), we have a solid-like response. Under this viewpoint, there is no fundamental difference between solids and liquids; it is only a matter of time scale. In the limit, when $\text{De} = 0$ one has a Newtonian liquid, and when $\text{De} = \infty$, an elastic solid.

---

[1] The terminology is due to M. Reiner.

4.2.2 Weissenberg Number

The Weissenberg number compares the elastic forces to the viscous effects. It has been variously defined, but usually as given in (4.12). Thus one can have a flow with a small Wi number and a large De number, and vice versa. We expect a significant non-Newtonian behaviour in a large Wi number flow, and therefore the constitutive equation must contain the relevant non-Newtonian physics. A different definition of the Weissenberg number is explored in Problem 5.3.

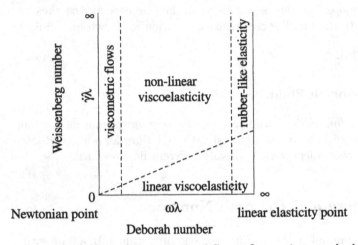

Fig. 4.8. Pipkin diagram delineates different flow regimes, and relevant constitutive equations

Pipkin's diagram (Pipkin and Tanner [11]) helps guide the choice of constitutive equations. In Fig. 4.8, the vertical axis represents the Weissenberg number and the horizontal axis the Deborah number. The Newtonian response is represented by a single point, at $De = 0 = Wi$. The elastic response is also represented by a single point, at $De = \infty$. Nearly steady flows, at low De numbers, can be analysed by viscometric motion (later), and fast flows, at large De numbers, a rubber-like response is expected. The region at low Wi numbers (at small strain amplitudes) can be handled by a linear viscoelastic model. The large domain in the middle, marked non-linear viscoelasticity, is the constitutive modeller's haven!

4.3 Some Guidelines in Constitutive Modelling

There are two alternatives for constitutive modelling: the continuum approach and the microstructure approach. In the continuum approach, the material is assumed to be a continuum, with no micro-inertial feature. The relevant variables are identified, and are related in a framework that ensured invariance under a change of frames. Different restrictions are then imposed to simplify the constitutive equation as far as practicable. In the microstructure approach, a physical model of the microstructure representing the material is postulated. Solving the deformation at that level using well-tested physical principles (Newton's laws, conservation laws, etc.) allows the average stress and strain to be related producing a constitutive equation. In the continuum approach one is usually left with a general constitutive equations, which may have some undetermined functions or functionals (loosely speaking, functionals are functions of functions). The details of these functions or functionals may be furnished by a relevant experiment. In the microstructure approach, the constitutive equations tend to be more specific and therefore more relevant to the material in question.

In the mid 50's, there were some intense activities in setting up a rigorous theoretical framework for continuum mechanics. Everything possible is set up in an axiomatic format. This has been good in focusing on what is permissible. However, it has the unfortunate consequence that it leaves the students with the impression that all one needs is a set of relevant variables and some principles – and that would allows to construct general constitutive equations for any material.

It is generally believed that relevant constitutive equations shlould be based on a (simplified) model of the microstructure. When the physics governing the microstructure interactions are complicated, one must not hesitate to introduce elements of continuum modelling, but the continuum approach should not completely replace the microstructure modelling.

4.3.1 Oldroyd Approach

The basic ideas behind constitutive modelling of finite deformation were well understood in the early 50's, but these ideas have not been extended to all continuous materials undergoing large deformation. It was Oldroyd[2] [12] who clearly enunciated that a constitutive equation must be based on

- The relative motion of the neighbourhood of a particle;
- The history of the metric tensor (i.e., strain tensor) associated with the particle;
- Convected coordinate system embedded in the material and deforming with it;
- Physical constants defining the symmetry of the material.

---

[2] James G. Oldroyd (1921–1982) was a Professor in Applied Mathematics at Universities of Wales and Liverpool. He made several important contributions to the constitutive equation formulation. The Oldroyd fluid was named after him.

It was unfortunate that his work has been grossly overlooked, see Tanner and Walters [13] for an interesting historical account. The later influential work of Noll [14] [15] put these ideas in an axiomatic form that is elegant and appealing to the generation of graduate students at that time. We will go through the principles as detailed by Noll, keeping in mind that models that are based on the microstructure tend to be more relevant.

4.3.2 Principle of Material Objectivity

Consider a change of frame

$$\mathbf{x} = \mathbf{c}\,(t) + \mathbf{Q}\,(t)\,\mathbf{x}, \tag{4.13}$$

which consists of a spatial translation (by $\mathbf{c}$) and rotation (through an orthogonal tensor $\mathbf{Q}$). A physical quantity is said to be objective, or frame-invariant, when it is invariant under the transformation (4.13). Specifically,

A scalar ϕ is invariant when its value remains unchanged under a change of frame (4.13)

$$\phi' = \phi. \tag{4.14}$$

A vector $\mathbf{u}$ is invariant when it transforms under a change of frame (4.13) according to

$$\mathbf{u}' = \mathbf{Q}\,(t)\,\mathbf{u}. \tag{4.15}$$

A tensor $\mathbf{T}$ is invariant when it transforms under a change of frame (4.13) according to

$$\mathbf{T}' = \mathbf{Q}\,(t)\,\mathbf{T}\mathbf{Q}\,(t)^{T}. \tag{4.16}$$

4.3.3 Objectivity of the Stress

The principle of material objectivity asserts that the stress tensor must be objective under a change of frame (4.13).

This principle pre-supposes that the material has no inertial feature at the microscale, that is, it is a continuous media. With micro-inertia, such as suspensions of microsized particles of a different density than the fluid, there will be a component of stress due to the particle inertia that is not objective. This is a "principle" for inertialess microstructure only. Indeed, if a microstructural model violates objectivity, the reason can always be found in the physics of the microstructure – if the physics are sound, there may be a very good reason the stress not being objective (Ryskin and Rallison [16]).

4.3.4 Frame Indifference

This principle enunciates that one does not obtain a new constitutive equation every time there is a change in frame of reference: the constitutive operator is the same for all observers in relative motion. The objectivity and frame indifference principles roughly correspond to Oldroyd's third point.

There are kinematical quantities that are not objective. We give a few examples here.

Deformation Gradient Tensor. Consider the deformation gradient tensor,

$$\mathbf{F}(t) = \left[\frac{\partial \mathbf{x}(t)}{\partial \mathbf{X}}\right]^T, \quad F_{ij} = \frac{\partial x_i}{\partial X_j}.$$

Under the change of frame (4.13),

$$F'_{ij} = \frac{\partial x'_i}{\partial X_j} = \frac{\partial x'_i}{\partial x_k}\frac{\partial x_k}{\partial X_j} = Q_{ik}F_{kj}, \quad \mathbf{F}' = \mathbf{QF}, \tag{4.17}$$

i.e., $\mathbf{F}$ is not frame-invariant. The relative deformation gradient, $\mathbf{F}_t(\tau) = \mathbf{F}(\tau)\mathbf{F}^{-1}(t)$, is not frame-invariant either,

$$\begin{aligned}
\mathbf{F}'_t(\tau) &= \mathbf{F}'(\tau)\mathbf{F}'(t)^{-1} = \mathbf{Q}(\tau)\mathbf{F}(\tau)\left[\mathbf{Q}(t)\mathbf{F}(t)\right]^{-1} \\
&= \mathbf{Q}(\tau)\mathbf{F}(\tau)\mathbf{F}(t)^{-1}\mathbf{Q}(t)^T \\
&= \mathbf{Q}(\tau)\mathbf{F}_t(\tau)\mathbf{Q}(t)^T.
\end{aligned} \tag{4.18}$$

Cauchy–Green Tensor. The left Cauchy–Green tensor is objective, but the right Cauchy–Green tensor is not:

$$\begin{aligned}
\mathbf{B}' &= \mathbf{F}'\mathbf{F}'^T = \mathbf{QF}(\mathbf{QF})^T = \mathbf{QFF}^T\mathbf{Q}^T = \mathbf{QBQ}^T, \\
\mathbf{C}' &= \mathbf{F}'^T\mathbf{F}' = (\mathbf{QF})^T\mathbf{QF} = \mathbf{FQ}^T\mathbf{QF} = \mathbf{C}.
\end{aligned} \tag{4.19}$$

The relative left Cauchy–Green tensors is not objective (note the argument of $\mathbf{Q}$):

$$\begin{aligned}
\mathbf{B}'_t(\tau) &= \mathbf{F}'_t(\tau)\mathbf{F}'_t(\tau)^T = \mathbf{Q}(\tau)\mathbf{F}_t(\tau)\mathbf{Q}(t)^T\left(\mathbf{Q}(\tau)\mathbf{F}_t(\tau)\mathbf{Q}(t)^T\right)^T \\
&= \mathbf{Q}(\tau)\mathbf{F}_t(\tau)\mathbf{Q}(t)^T\mathbf{Q}(t)\mathbf{F}_t(\tau)^T\mathbf{Q}(\tau)^T \\
&= \mathbf{Q}(\tau)\mathbf{B}_t(\tau)\mathbf{Q}(\tau)^T,
\end{aligned} \tag{4.20}$$

but the right Cauchy–Green tensor is,

$$\begin{aligned}
\mathbf{C}'_t(\tau) &= \mathbf{F}'_t(\tau)^T\mathbf{F}'_t(\tau) = \left(\mathbf{Q}(\tau)\mathbf{F}_t(\tau)\mathbf{Q}(t)^T\right)^T\mathbf{Q}(\tau)\mathbf{F}_t(\tau)\mathbf{Q}(t)^T \\
&= \mathbf{Q}(t)\mathbf{F}_t(\tau)^T\mathbf{Q}(\tau)^T\mathbf{Q}(\tau)\mathbf{F}_t(\tau)\mathbf{Q}(t) \\
&= \mathbf{Q}(t)\mathbf{C}_t(\tau)\mathbf{Q}(t)^T.
\end{aligned} \tag{4.21}$$

Velocity Gradient. The velocity gradient is not objective. To see this, we note that the motion is transformed according to,

$$\mathbf{M}'(\mathbf{X}, t) = \mathbf{c}(t) + \mathbf{Q}(t)\mathbf{M}(\mathbf{X}, t),$$ (4.22)

since the motion is given by $\mathbf{x}(t) = \mathbf{M}(\mathbf{X}, t)$. Thus the velocity transforms according to

$$\hat{\mathbf{u}}'(\mathbf{X}, t) = \dot{\mathbf{c}}(t) + \mathbf{Q}(t)\hat{\mathbf{u}}(\mathbf{X}, t) + \dot{\mathbf{Q}}(t)\mathbf{x}(\mathbf{X}, t).$$ (4.23)

Expressing this in the Eulerian sense

$$\mathbf{u}'(\mathbf{x}, t) = \dot{\mathbf{c}}(t) + \mathbf{Q}(t)\mathbf{u}(\mathbf{x}, t) + \dot{\mathbf{Q}}(t)\mathbf{x}.$$ (4.24)

The velocity gradient thus transforms accordingly

$$L'_{ij} = \frac{\partial u'_i}{\partial x'_j} = \frac{\partial u'_i}{\partial x_k}\frac{\partial x_k}{\partial x'_j} = \left(Q_{il}\frac{\partial u_l}{\partial x_k} + \dot{Q}_{ik}\right)Q^T_{kj},$$

that is,

$$\mathbf{L}' = \mathbf{Q}\mathbf{L}\mathbf{Q}^T + \dot{\mathbf{Q}}\mathbf{Q}^T.$$ (4.25)

$\dot{\mathbf{Q}}\mathbf{Q}^T$ is anti-symmetry (why?), and thus the symmetric part of (4.25), or the strain rate tensor, is objective while the anti-symmetric part of (4.25), or the vorticity tensor, is not:

$$\begin{aligned}\mathbf{D}' &= \mathbf{Q}\mathbf{D}\mathbf{Q}^T,\\ \mathbf{W}' &= \mathbf{Q}\mathbf{W}\mathbf{Q}^T + \dot{\mathbf{Q}}\mathbf{Q}^T.\end{aligned}$$ (4.26)

Rivlin–Ericksen Tensors. All the Rivlin–Ericksen tensors are objective. We have just seen that the first Rivlin–Ericksen tensor is objective,

$$\mathbf{A}'_1 = \mathbf{Q}\mathbf{A}_1\mathbf{Q}^T.$$ (4.27)

Let all the Rivlin–Ericksen tensors up to order $1 \le n$ be objective. The next Rivlin–Ericksen tensor of order $n + 1$ transforms according to

$$\begin{aligned}\mathbf{A}'_{n+1} &= \frac{d}{dt}\mathbf{A}'_n + \mathbf{A}'_n\mathbf{L}' + \mathbf{L}'^T\mathbf{A}'_n\\ &= \frac{d}{dt}\left(\mathbf{Q}\mathbf{A}_n\mathbf{Q}^T\right) + \mathbf{Q}\mathbf{A}_n\mathbf{Q}^T\left(\mathbf{Q}\mathbf{L}\mathbf{Q}^T + \dot{\mathbf{Q}}\mathbf{Q}^T\right)\\ &\quad + \left(\mathbf{Q}\mathbf{L}\mathbf{Q}^T + \dot{\mathbf{Q}}\mathbf{Q}^T\right)^T\mathbf{Q}\mathbf{A}_n\mathbf{Q}^T\\ &= \mathbf{Q}\left[\frac{d}{dt}\mathbf{A}_n + \mathbf{A}_n\mathbf{L} + \mathbf{L}^T\mathbf{A}_n\right]\mathbf{Q}^T\\ &\quad + \dot{\mathbf{Q}}\mathbf{A}_n\mathbf{Q}^T + \mathbf{Q}\mathbf{A}_n\dot{\mathbf{Q}}^T + \mathbf{Q}\mathbf{A}_n\mathbf{Q}^T\dot{\mathbf{Q}}\mathbf{Q}^T + \mathbf{Q}\dot{\mathbf{Q}}^T\mathbf{Q}\mathbf{A}_n\mathbf{Q}^T.\end{aligned}$$

Since $\mathbf{Q}\mathbf{Q}^T = \mathbf{I}$,

$$\dot{\mathbf{Q}} = -\mathbf{Q}\dot{\mathbf{Q}}^T\mathbf{Q}, \quad \dot{\mathbf{Q}}^T = -\mathbf{Q}^T\dot{\mathbf{Q}}\mathbf{Q}^T,$$

and we find that

$$\mathbf{A}'_{n+1} = \mathbf{Q}\left[\frac{d}{dt}\mathbf{A}_n + \mathbf{A}_n\mathbf{L} + \mathbf{L}^T\mathbf{A}_n\right]\mathbf{Q}^T, \tag{4.28}$$

i.e., the $(n+1)$th-order Rivlin–Ericksen tensor is also objective. By the process of induction, all the Rivlin–Ericksen tensors are objective. The next two principles enunciate our common sense about material behaviour.

4.3.5 Principle of Local Action

The principle of local action embodies the idea that only particles near a point should be involved in determining the stress at that point. This is consistent with the exclusion of long-range forces, which have already been included in body forces. This is Oldroyd's first point.

4.3.6 Principle of Determinism

This principle states the obvious, that the current stress state in the material is determined by the past history of the motion. Future state of the motion has no say in the current state of the stress; i.e., the material possesses no clairvoyance.

In addition to these principles, there may be restrictions imposed on the constitutive equation because of the symmetry of the material. These symmetry restrictions are discussed separately. This is satisfied by Oldroyd second and third points.

4.4 Integrity Bases

4.4.1 Isotropic Scalar-Valued Functions

Consider a scalar-valued function of a vector $\mathbf{u}$, and suppose that this function satisfies

$$f(\mathbf{u}) = f(\mathbf{Q}\mathbf{u}), \tag{4.29}$$

for every orthogonal tensor $\mathbf{Q}$. Such a function is called *isotropic*. We are interested in how f depends on $\mathbf{u}$. The answer is simple because only the magnitude of $\mathbf{u}$ is invariant under every orthogonal tensor. Thus f must be a function of $u = |\mathbf{u}|$. In that sense, $u = |\mathbf{u}|$ is called an *integrity basis*.

Likewise, a scalar-valued function of two vectors $\mathbf{u}$ and $\mathbf{v}$ satisfying

$$f(\mathbf{u}, \mathbf{v}) = f(\mathbf{Qu}, \mathbf{Qv}), \qquad (4.30)$$

for every orthogonal $\mathbf{Q}$, is called an isotropic function of its two arguments. Since

$$\mathbf{u} \cdot \mathbf{u}, \quad \mathbf{v} \cdot \mathbf{v}, \quad \mathbf{u} \cdot \mathbf{v} \qquad (4.31)$$

are the only three invariants under rotation (their lengths and the angle between them are invariant, Weyl [17]),

$$f(\mathbf{u}, \mathbf{v}) = f(\mathbf{u} \cdot \mathbf{u}, \mathbf{v} \cdot \mathbf{v}, \mathbf{u} \cdot \mathbf{v}). \qquad (4.32)$$

The three scalar invariants form the integrity basis for $f(\mathbf{u}, \mathbf{v})$.

Similarly, f is said to be an isotropic function of a tensor $\mathbf{S}$ if for every orthogonal $\mathbf{Q}$,

$$f(\mathbf{S}) = f\left(\mathbf{QSQ}^T\right). \qquad (4.33)$$

Here, f must be a function of the three invariants of $\mathbf{S}$,

$$\operatorname{tr}\mathbf{S}, \quad \operatorname{tr}\mathbf{S}^2, \quad \operatorname{tr}\mathbf{S}^3, \qquad (4.34)$$

or, equivalently, of the three eigenvalues of $\mathbf{S}$. These three invariants form the integrity bases for $f(\mathbf{S})$.

The invariants that can be formed from two tensors $\mathbf{A}$ and $\mathbf{B}$ are

$$\begin{array}{c} \operatorname{tr}\mathbf{A}, \quad \operatorname{tr}\mathbf{A}^2, \quad \operatorname{tr}\mathbf{A}^3, \quad \operatorname{tr}\mathbf{B}, \quad \operatorname{tr}\mathbf{B}^2, \quad \operatorname{tr}\mathbf{B}^3, \\ \operatorname{tr}\mathbf{AB}, \quad \operatorname{tr}\mathbf{A}^2\mathbf{B}, \quad \operatorname{tr}\mathbf{AB}^2, \quad \operatorname{tr}\mathbf{A}^2\mathbf{B}^2. \end{array} \qquad (4.35)$$

Thus, an isotropic scalar-valued function of $\mathbf{A}$ and $\mathbf{B}$ must be a function of these 10 invariants. These form the integrity basis for $f(\mathbf{A}, \mathbf{B})$.

We list another integrity basis for a scalar valued, isotropic function of a symmetric second order tensor $\mathbf{S}$ and two vectors $\mathbf{u}$ and $\mathbf{v}$:

$$\begin{array}{l} \operatorname{tr}\mathbf{S}, \ \operatorname{tr}\mathbf{S}^2, \ \operatorname{tr}\mathbf{S}^3, \ \mathbf{u}\cdot\mathbf{u}, \ \mathbf{u}\cdot\mathbf{v}, \ \mathbf{v}\cdot\mathbf{v}, \ \mathbf{u}\cdot\mathbf{S}\mathbf{u}, \ \mathbf{u}\cdot\mathbf{S}^2\mathbf{u}, \\ \mathbf{v}\cdot\mathbf{S}\mathbf{v}, \ \mathbf{v}\cdot\mathbf{S}^2\mathbf{v}, \ \mathbf{u}\cdot\mathbf{S}\mathbf{v}, \ \mathbf{u}\cdot\mathbf{S}^2\mathbf{v}. \end{array} \qquad (4.36)$$

4.4.2 Isotropic Vector-Valued Functions

Suppose that $\mathbf{w} = \mathbf{g}(\mathbf{v})$ is a vector-valued function of the vector $\mathbf{u}$. It is called isotropic if, for every orthogonal tensor $\mathbf{Q}$,

$$\mathbf{Q}\mathbf{g}(\mathbf{v}) = \mathbf{g}(\mathbf{Qv}). \qquad (4.37)$$

Now, define a scalar-valued function of two vectors $\mathbf{u}$ and $\mathbf{v}$ through

$$f(\mathbf{u}, \mathbf{v}) = \mathbf{u} \cdot \mathbf{g}(\mathbf{v}).$$

Thus, since $\mathbf{g}$ is isotropic,

$$f\left(\mathbf{Qu}, \mathbf{Qv}\right) = \mathbf{Qu} \cdot \mathbf{g}\left(\mathbf{Qv}\right) = \mathbf{Qu} \cdot \mathbf{Qg}\left(\mathbf{v}\right) = \mathbf{u} \cdot \mathbf{g}\left(\mathbf{v}\right),$$

since $\mathbf{Qu} \cdot \mathbf{Qv} = Q_{ij} u_j Q_{ik} v_k = \delta_{jk} u_j v_k = \mathbf{u} \cdot \mathbf{v}$. Consequently, f is an isotropic function of its two arguments, and thus it is a function of the invariants listed in (4.31). However, by its definition, $f(\mathbf{u}, \mathbf{v})$ is linear in its first argument, and therefore

$$f\left(\mathbf{u}, \mathbf{v}\right) = \mathbf{u} \cdot h\left(\mathbf{v} \cdot \mathbf{v}\right) \mathbf{v}.$$

It follows that

$$\mathbf{g}\left(\mathbf{v}\right) = h\left(\mathbf{v} \cdot \mathbf{v}\right) \mathbf{v}. \tag{4.38}$$

4.4.3 Isotropic Tensor-Valued Functions

A symmetric tensor-valued function of a symmetric tensor $\mathbf{B}$ is isotropic if

$$\mathbf{QG}\left(\mathbf{B}\right) \mathbf{Q}^T = \mathbf{G}\left(\mathbf{QBQ}^T\right), \tag{4.39}$$

for every orthogonal tensor $\mathbf{Q}$.

Now, define a scalar function of two symmetric tensors through

$$f\left(\mathbf{A}, \mathbf{B}\right) = \operatorname{tr}\left[\mathbf{AG}\left(\mathbf{B}\right)\right].$$

From its definition,

$$
\begin{aligned}
f\left(\mathbf{QAQ}^T, \mathbf{QBQ}^T\right) &= \operatorname{tr}\left[\mathbf{QAQ}^T \mathbf{G}\left(\mathbf{QBQ}^T\right)\right] \\
&= \operatorname{tr}\left[\mathbf{QAQ}^T \mathbf{QG}\left(\mathbf{B}\right) \mathbf{Q}^T\right] \\
&= \operatorname{tr}\left[\mathbf{QAG}\left(\mathbf{B}\right) \mathbf{Q}^T\right] = Q_{ij} A_{jk} G_{kl} Q_{il} \\
&= \operatorname{tr}\left[\mathbf{AG}\left(\mathbf{B}\right)\right].
\end{aligned}
$$

That is f is isotropic in its two arguments. It is therefore a function of the ten invariants listed in (4.35). Since $f(\mathbf{A}, \mathbf{B})$ is linear in its first argument,

$$f\left(\mathbf{A}, \mathbf{B}\right) = \operatorname{tr}\left[\mathbf{A}\left(g_0 \mathbf{I} + g_1 \mathbf{B} + g_2 \mathbf{B}^2\right)\right],$$

and consequently

$$\mathbf{G}\left(\mathbf{B}\right) = g_0 \mathbf{I} + g_1 \mathbf{B} + g_2 \mathbf{B}^2, \tag{4.40}$$

where g_0, g_1, g_2 are scalar-valued functions of the three invariants of $\mathbf{B}$.

The underlying principle is (Pipkin and Rivlin [18]): to find the form for isotropic vector-valued, or symmetric tensor-valued functions of vector or symmetric vector, first form an artificial scalar product with a second vector or another symmetric tensor, and then find the relevant integrity bases for this isotropic scalar-valued function. Finally, because of the linearity in its definition, non-linear terms can be discarded, arriving at the correct form for the original isotropic function. For functions which are isotropic, or transversely isotropic or have crystal classes as their symmetric groups, see the review article by Spencer [19]. For functions which are invariant under the full unimodular group, see Fahy and Smith [20].

4.5 Symmetry Restrictions

4.5.1 Unimodular Matrix

Let X and X' be two adjacent particles with positions $\mathbf{X}$ and $\mathbf{X} + d\mathbf{X}$ in the reference configuration. Now consider a change in the local configuration so that X remains at $\mathbf{X}$, while X' goes to $\mathbf{X} + d\mathbf{X}'$. We assume that the gradient of this transformation is $\mathbf{H}$, where

$$d\mathbf{X}' = \mathbf{H}d\mathbf{X}, \tag{4.41}$$

and $\mathbf{H}$ is a proper unimodular matrix, i.e., $\det \mathbf{H} = 1$ (the configuration change should not lead to a change in volume).

Now if the motion is such that X goes to $\mathbf{x}$, and X' goes to $\mathbf{x} + d\mathbf{x}$, then we have a new motion M':

$$\begin{aligned}
M' &: X \text{ or } \mathbf{X} \mapsto \mathbf{x}, \; X' \text{ or } \mathbf{X} + d\mathbf{X}' \mapsto \mathbf{x} + d\mathbf{x}, \\
M &: X \text{ or } \mathbf{X} \mapsto \mathbf{x}, \; X \text{ or } \mathbf{X} + d\mathbf{X} \mapsto \mathbf{x} + d\mathbf{x}.
\end{aligned} \tag{4.42}$$

The two mappings are different, because the two shapes about $\mathbf{X}$ are mapped into the same shape about $\mathbf{x}$ (see Fig. 4.9). Now, since

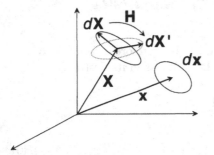

Fig. 4.9. A change in the local configuration leads to the same shape after deformation

$$dx = \mathbf{F}d\mathbf{X} = \mathbf{F}'d\mathbf{X}' = \mathbf{F}'\mathbf{H}d\mathbf{X},$$

we have

$$\mathbf{F}'\mathbf{H} = \mathbf{F} \quad \therefore \quad \mathbf{F}' = \mathbf{F}\mathbf{H}^{-1}. \tag{4.43}$$

The strain measures can be calculated

$$\mathbf{B}' = \mathbf{F}'\mathbf{F}'^{T} = \mathbf{F}\mathbf{H}^{-1}\mathbf{H}^{-T}\mathbf{F}^{T}, \tag{4.44}$$

$$\mathbf{C}' = \mathbf{F}'^{T}\mathbf{F}' = \mathbf{H}^{-T}\mathbf{F}^{T}\mathbf{F}\mathbf{H}^{-1} = \mathbf{H}^{-T}\mathbf{C}\mathbf{H}^{-1}, \tag{4.45}$$

$$\mathbf{F}'_{t}\left(\tau\right) = \mathbf{F}'\left(\tau\right)\mathbf{F}'\left(t\right)^{-1} = \mathbf{F}\left(\tau\right)\mathbf{H}^{-1}\mathbf{H}\mathbf{F}\left(t\right)^{-1} = \mathbf{F}_{t}\left(\tau\right), \tag{4.46}$$

$$\mathbf{C}'_{t}\left(\tau\right) = \mathbf{F}'_{t}\left(\tau\right)^{T}\mathbf{F}'_{t}\left(\tau\right) = \mathbf{C}_{t}\left(\tau\right). \tag{4.47}$$

The relative strain measure $\mathbf{C}_{t}\left(\tau\right)$ is not sensitive to the unimodular changes about X.

4.5.2 Symmetry Group

Suppose that we are interested in certain constitutive property $\wp$, say the stress tensor, that depends on the kinematics. Moreover, suppose that some unimodular changes of the local shape about X leave this quantity unchanges. Let

$$\mathcal{G}_{\wp} = \{\mathbf{I}, \mathbf{H}_{1}, \mathbf{H}_{2}, \ldots\} \tag{4.48}$$

be the set of all the unimodular transformations that preserves $\wp$ then $\mathcal{G}_{\wp}$ is a group, called the $\wp$-symmetry group, i.e., the group of unimodular changes in the neighbourhood of X that leaves $\wp$ invariant.

4.5.3 Isotropic Materials

Different materials have different symmetry groups: there are isotropic, transversely isotropic groups, etc. We are mainly concerned with isotropic materials, where the symmetry group is the proper orthogonal group, $\mathbf{H}^{-1} = \mathbf{H}^{T}$.

4.6 Isotropic Elastic Materials

An elastic material is one in which the stress is a function of the deformation gradient:

$$\mathbf{T} = \mathbf{f}\left(\mathbf{F}\right), \tag{4.49}$$

where $\mathbf{f}$ is a symmetric tensor-valued function of $\mathbf{F}$. For an isotropic material, the symmetry group is the full proper orthogonal group, that is, for every orthogonal $\mathbf{H}$, we have

$$\mathbf{f}\left(\mathbf{F}\right) = \mathbf{f}\left(\mathbf{FH}\right) \tag{4.50}$$

Isotropic Constraint. Now $\mathbf{F}$ has the unique polar decomposition

$$\mathbf{F} = \mathbf{VR},$$

where $\mathbf{R}$ is orthogonal. Thus

$$\mathbf{F}' = \mathbf{FH} = \mathbf{VRH},$$

and because $\mathbf{H}$ is orthogonal, $\mathbf{RH}$ is orthogonal. In another words, $\mathbf{F}'$ can be chosen in the set $\{\mathbf{V}, \mathbf{VQ}_1, \mathbf{VQ}_2, \ldots\}$, where $\mathbf{Q}_i$ are orthogonal. Thus $\mathbf{f}$ is a function of $\mathbf{V}$ alone. Since $\mathbf{B} = \mathbf{FF}^T = \mathbf{V}^2$, $\mathbf{f}$ is a function of the strain $\mathbf{B}$:

$$\mathbf{T} = \mathbf{f}(\mathbf{F}) = \mathbf{f}(\mathbf{V}) = \mathbf{f}(\mathbf{B}). \tag{4.51}$$

Objectivity Constraint. Objectivity imposes the following constraint on $\mathbf{T}$:

$$\mathbf{T}' = \mathbf{QTQ}^T, \tag{4.52}$$

every orthogonal $\mathbf{Q}$. Since $\mathbf{T}' = \mathbf{f}(\mathbf{B}') = \mathbf{f}\left(\mathbf{QBQ}^T\right)$ (frame indifference of the stress, objectivity of $\mathbf{B}$) the requirement (4.52) becomes

$$\mathbf{Qf}(\mathbf{B})\mathbf{Q}^T = \mathbf{f}\left(\mathbf{QBQ}^T\right). \tag{4.53}$$

Thus $\mathbf{f}$ is an isotropic function of $\mathbf{B}$ (see (4.39)). The general form for $\mathbf{f}$ has been found in (4.40), thus

$$\mathbf{f}(\mathbf{B}) = \alpha_0\mathbf{I} + \alpha_1\mathbf{B} + \alpha_2\mathbf{B}^2, \tag{4.54}$$

where the scalar coefficients are functions of the three invariants of $\mathbf{B}$. Alternatively, one may use the Cayley–Hamilton theorem and express $\mathbf{B}^2$ in terms of $\mathbf{B}$ and $\mathbf{B}^{-1}$.

Mooney and neo-Hookean Materials. Therefore, the general constitutive equation for an isotropic elastic solid is given by

$$\mathbf{T} = \beta_0\mathbf{I} + \beta_1\mathbf{B} - \beta_2\mathbf{B}^{-1}, \tag{4.55}$$

where β_i are functions of the three invariants of $\mathbf{B}$. The term $\beta_0\mathbf{I}$ can be absorbed in the hydrostatic pressure. The case when β_1, β_2 are constant is called the *Mooney material* [21] or sometimes the Mooney–Rivlin material, in deference to Rivlin. In addition, if $\beta_2 = 0$ it is called the *neo-Hookean material* [22].

If the material behaves like a Mooney model, a plot of $T_{ZZ}/(\lambda - \lambda^2)$ against λ^{-1} should be a straigth line, with slope β_2 and intercept β_1. The data from Rivlin and Saunder (see [22]) showed that the Mooney model is inadequate: in compression, their data indicated that $\beta_2 \approx 0$, whereas in tension, their data showed that β_2/β_1 varies from 0.3 to 1. However, the Mooney model should be adequate for most qualitative purposes.

4.7 The Simple Material

Noll [15] defined a simple material (solid or liquid) as one in which the current stress is a functional (function of function) of the history of the deformation gradient $\mathbf{F}(\tau)$, $-\infty < \tau \leq t$:

$$\mathbf{T}(t) = \mathcal{G}(\mathbf{F}(\tau)), \quad -\infty < \tau \leq t. \tag{4.56}$$

Objectivity and frame indifference require,

$$\mathbf{Q}(t)\,\mathcal{G}(\mathbf{F}(\tau))\,\mathbf{Q}(t)^T = \mathcal{G}(\mathbf{Q}(\tau)\,\mathbf{F}(\tau)), \tag{4.57}$$

for all rotational histories $\mathbf{Q}(\tau)$, $-\infty < \tau \leq t$.

Recall now the polar decomposition for $\mathbf{F}$, $\mathbf{F}(\tau) = \mathbf{R}(\tau)\,\mathbf{U}(\tau)$. Since $\mathbf{Q}$ is arbitrary, we choose $\mathbf{Q}(\tau) = \mathbf{R}(\tau)^T$. Thus (4.57) requires

$$\mathcal{G}(\mathbf{F}(\tau)) = \mathbf{R}(t)\,\mathcal{G}(\mathbf{U}(\tau))\,\mathbf{R}(t)^T$$
$$= \mathbf{F}(t)\,\mathbf{U}(t)^{-1}\,\mathcal{G}(\mathbf{U}(\tau))\,\mathbf{U}(t)^{-1}\,\mathbf{F}(t)^T.$$

Using the definition $\mathbf{C} = \mathbf{U}^2$, we define a new functional through

$$\mathcal{F}(\mathbf{C}(\tau)) = \mathbf{U}(t)^{-1}\,\mathcal{G}(\mathbf{U}(\tau))\,\mathbf{U}(t)^{-1}. \tag{4.58}$$

Thus for a simple material,

$$\mathbf{T}(t) = \mathbf{F}(t)\,\mathcal{F}(\mathbf{C}(\tau))\,\mathbf{F}(t)^T, \quad -\infty < \tau \leq t. \tag{4.59}$$

In addition, from (4.104),

$$\mathbf{T}(t) = \mathcal{G}(\mathbf{F}(\tau)) = \mathbf{F}(t)\,\mathcal{F}\left(\mathbf{F}(t)^T\,\mathbf{C}_t(\tau)\,\mathbf{F}(t)\right)\mathbf{F}(t)^T, \quad -\infty < \tau \leq t. \tag{4.60}$$

This says that the current stress is a functional of the history of the right relative Cauchy–Green tensor, and the current value of the deformation gradient. That is, we may define a new functional

$$\mathcal{G}(\mathbf{F}(\tau)) = \mathcal{H}(\mathbf{C}_t(\tau), \mathbf{F}(t)). \tag{4.61}$$

Replacing $\mathbf{F}(\tau)$ by $\mathbf{F}(\tau)\mathbf{H}$, where $\mathbf{H}$ is unimodular, leaves $\mathbf{C}_t(\tau)$ unchanged, and thus,

$$\mathcal{H}(\mathbf{C}_t(\tau), \mathbf{F}(t)) = \mathcal{H}(\mathbf{C}_t(\tau), \mathbf{F}(t)\mathbf{H}). \tag{4.62}$$

In addition, this functional must obeys the objectivity restriction (4.57):

$$\mathbf{Q}(t)\,\mathcal{H}(\mathbf{C}_t(\tau), \mathbf{F}(t))\,\mathbf{Q}(t)^T = \mathcal{H}\left(\mathbf{Q}(t)\,\mathbf{C}_t(\tau)\,\mathbf{Q}(t)^T, \mathbf{Q}(t)\,\mathbf{F}(t)\right) \tag{4.63}$$

noting the objectivity of the relative right Cauchy–Green strain tensor (4.21).

4.7.1 Simple Fluid

If the material is an isotropic fluid (note that fluids are isotropic in Noll's definition), the stress is invariant under the full orthogonal group. Noll showed in this case, the current stress is given by

$$\mathbf{T}(t) = \mathcal{F}\left(\mathbf{C}_t(\tau), \rho(t)\right), \quad -\infty < \tau \le t. \tag{4.64}$$

This functional must satisfy objectivity,

$$\mathbf{Q}(t)\,\mathcal{F}\left(\mathbf{C}_t(\tau), \rho(t)\right)\mathbf{Q}(t)^T = \mathcal{F}\left(\mathbf{Q}(t)\,\mathbf{C}_t(\tau)\,\mathbf{Q}(t)^T, \rho(t)\right), \tag{4.65}$$

for all orthogonal tensor $\mathbf{Q}$.

4.7.2 Incompressible Simple Fluid

Incompressibility has been introduced as a simplification of the real material behavior. This demands that

$$\rho(t) = \rho_R, \quad \det \mathbf{F}(t) = 1, \quad \det \mathbf{C}(t) = 1, \quad \nabla \cdot \mathbf{u} = 0. \tag{4.66}$$

Its adoption implies that the constitutive relation can only determine the stress up to an isotropic part (the hydrostatic pressure); this hydrostatic pressure must be determined by the equations of balance. We write

$$\begin{aligned}\mathbf{T} &= -P\mathbf{I} + \mathbf{S}, \\ \mathbf{S}(t) &= \mathcal{F}\left(\mathbf{C}_t(\tau)\right), \quad -\infty < \tau \le t.\end{aligned} \tag{4.67}$$

$\mathbf{S}$ is called the extra stress. Of course the functional $\mathcal{F}$ must satisfy objectivity:

$$\mathbf{Q}(t)\,\mathcal{F}\left(\mathbf{C}_t(\tau)\right)\mathbf{Q}(t)^T = \mathcal{F}\left(\mathbf{Q}(t)\,\mathbf{C}_t(\tau)\,\mathbf{Q}(t)^T\right). \tag{4.68}$$

4.7.3 Fading Memory

The idea of fading memory embodies the notion that distant events (large τ) should have less bearing on the current stress than recent past. This idea can be implemented in the functional in various ways, through integral or differential operators. We close this chapter with two classes of constitutive relations obtained by assuming that the fluid memory is instantaneous, and that the deformation is small in some sense, i.e., the relative strain tensor hardly departs from the unit tensor.

4.8 Order Fluids

When the fluid memory is catastrophic (i.e., instantaneous), we can assume that that stress is an isotropic function of the Rivlin–Ericksen tensors $\mathbf{A}_n$. Recall that the relative strain tensor can be expressed as a Taylor series with coefficients $\mathbf{A}_n$,

$$\mathbf{S} = \mathbf{f}\left(\mathbf{A}_1, \mathbf{A}_2, \dots, \mathbf{A}_N\right). \tag{4.69}$$

The physical dimensions of $\mathbf{A}_n$ are T^{-n}, where T is the time. A sequence of approximations to $\mathbf{f}$, correct to order T^n, $n = 1, 2, \dots$ can be developed. To first order, the Newtonian fluid:

$$\mathbf{S} = \mathbf{S}^{(1)} = \eta_0 \mathbf{A}_1. \tag{4.70}$$

To second order, the second-order fluid model:

$$\mathbf{S} = \mathbf{S}^{(2)} = \mathbf{S}^{(1)} + \left(\nu_1 + \nu_2\right)\mathbf{A}_1^2 - \frac{\nu_1}{2}\mathbf{A}_2. \tag{4.71}$$

To third order, the third-order fluid model:

$$\mathbf{S} = \mathbf{S}^{(3)} = \mathbf{S}^{(2)} + \alpha_0 \left(\operatorname{tr}\mathbf{A}_1^2\right)\mathbf{A}_1 + \alpha_1 \left(\mathbf{A}_1\mathbf{A}_2 + \mathbf{A}_2\mathbf{A}_1\right) + \alpha_3 \mathbf{A}_3. \tag{4.72}$$

Higher-order fluids can be developed in the same manner. These order fluids possess no memory, and using them to describe to describe memory phenomena may lead to disaster.

4.8.1 Unsteady Motion

To see that the order fluids are unsuitable for discussing unsteady motion, we consider the flow in a channel of width h,

$$u = u\left(y, t\right), \quad v = 0, \quad w = 0. \tag{4.73}$$

The first Rivlin–Ericksen tensor and its square are given by

$$[\mathbf{A}_1] = \begin{bmatrix} 0 & \partial u/\partial y & 0 \\ \partial u/\partial y & 0 & 0 \\ 0 & 0 & 0 \end{bmatrix}, \quad [\mathbf{A}_1^2] = \begin{bmatrix} (\partial u/\partial y)^2 & 0 & 0 \\ 0 & (\partial u/\partial y)^2 & 0 \\ 0 & 0 & 0 \end{bmatrix}.$$

The second Rivlin–Ericksen tensor is

$$\begin{aligned} [\mathbf{A}_2] &= \left[\dot{\mathbf{A}}_1\right] + [\mathbf{A}_1\mathbf{L}] + [\mathbf{L}^T\mathbf{A}_1] \\ &= \begin{bmatrix} 0 & \partial^2 u/\partial t \partial y & 0 \\ \partial^2 u/\partial t \partial y & 0 & 0 \\ 0 & 0 & 0 \end{bmatrix} + \begin{bmatrix} 0 & 0 & 0 \\ 0 & 2\left(\partial u/\partial y\right)^2 & 0 \\ 0 & 0 & 0 \end{bmatrix}. \end{aligned}$$

The stress components are

$$S_{xy} = \eta_0 \frac{\partial u}{\partial y} - \frac{\nu_1}{2}\frac{\partial^2 u}{\partial t \partial y}, \quad S_{xx} = (\nu_1 + \nu_2)\left(\frac{\partial u}{\partial y}\right)^2,$$

$$S_{yy} = \nu_2 \left(\frac{\partial u}{\partial y}\right)^2, \quad S_{zz} = 0.$$

Now, the equations of balance read

$$\rho\frac{\partial u}{\partial t} = -\frac{\partial P}{\partial x} + \frac{\partial S_{xx}}{\partial x} + \frac{\partial S_{xy}}{\partial y} + \frac{\partial S_{xz}}{\partial z},$$

$$0 = -\frac{\partial P}{\partial y} + \frac{\partial S_{yx}}{\partial x} + \frac{\partial S_{yy}}{\partial y} + \frac{\partial S_{yz}}{\partial z},$$

$$0 = -\frac{\partial P}{\partial z} + \frac{\partial S_{zx}}{\partial x} + \frac{\partial S_{zy}}{\partial y} + \frac{\partial S_{zz}}{\partial z}.$$

With no pressure gradient, the equations of motion reduce to

$$\rho\frac{\partial u}{\partial t} = \eta\frac{\partial^2 u}{\partial y^2} - \frac{\nu_1}{2}\frac{\partial^3 u}{\partial t \partial y^2}, \quad u(0,t) = 0, \ u(h,t) = 0. \qquad (4.74)$$

This is a linear third-order partial differential; we look for a solution by separation of variables:

$$u(y,t) = \sum_{n=1} \phi_n(t)\psi_n(y),$$

where

$$\rho\dot{\phi}_n\psi_n = \eta_0\phi_n\psi_n'' - \frac{1}{2}\nu_1\dot{\phi}_n\psi_n''.$$

We can try the Fourier series for ψ_n:

$$\psi_n = a_n \sin\frac{n\pi y}{h},$$

which leads to

$$\left[\rho - \frac{1}{2}\nu_1\frac{n^2\pi^2}{h^2}\right]\dot{\phi}_n = -\eta_0\frac{n^2\pi^2}{h^2}\phi_n.$$

This has the solution

$$\phi_n(t) = \phi_n^0 \exp\left(\frac{\eta_0 t}{\nu_1/2 - \rho h^2/n^2\pi^2}\right). \qquad (4.75)$$

Clearly, for given material properties and geometry, we can always find n so that the exponent is positive (a positive exponent implies instability: the solution is unbounded in time); that is the solution (4.75) is unstable to any disturbance of the form stated. Only when $\nu_1 = 0$ (Newtonian fluid) that the solution (4.75) is stable. Thus all unsteady flows are too fast for the second-order fluid to handle (and indeed for all order fluids).

4.8.2 Velocity Field in a Second-Order Fluid

The Newtonian pressure field p_N has to obey

$$\nabla p_N = \eta_0 \nabla \cdot \mathbf{A}_1 - \rho \mathbf{a}, \tag{4.76}$$

where $\mathbf{a}$ is the acceleration field. If the same velocity field were to occur in the second-order fluid, then an additional pressure term p_S will arise and this has to satisfy

$$\nabla p_S = \nabla \cdot \left[(\nu_1 + \nu_2)\mathbf{A}_1^2 - \frac{1}{2}\nu_1 \mathbf{A}_2 \right]. \tag{4.77}$$

There are a few special classes of flows that the right side of (4.77) is a scalar. We will consider two special cases.

Potential Flows. When the velocity field is a potential flow and, consequently $\mathbf{u} = \nabla\phi$, incompressibility demands that $\phi_{,ii} = 0$. Since

$$(\mathbf{A}_1)_{ij} = 2\phi_{,ij}, \tag{4.78}$$

it follows that

$$\nabla \cdot \mathbf{A}_1 = \mathbf{0}. \tag{4.79}$$

Hence, the Newtonian pressure field is given by

$$p_N = -\rho \left[(\phi_{,t} + \frac{1}{2}\mathbf{u} \cdot \mathbf{u} \right]. \tag{4.80}$$

The results from Problem 4.10, (4.111), show clearly that in potential flows, the Newtonian and second-order fluid velocity fields are identical, with the second-order pressure term given by

$$p_S = \frac{1}{8} (\nu_1 + 4\nu_2) \operatorname{tr} \mathbf{A}_1^2. \tag{4.81}$$

Plane Creeping Flows. The second case where the Newtonian velocity field is also the second-order fluid velocity field is the steady plane flow, where the velocity takes the form $\mathbf{u} = u(x,y)\mathbf{i} + v(x,y)\mathbf{j}$. The Cayley–Hamilton theorem and incompressibility together show that

$$\mathbf{A}_1^2 + (\det \mathbf{A}_1)\mathbf{1} = \mathbf{0}. \tag{4.82}$$

Thus,

$$\det \mathbf{A}_1 = -\frac{1}{2} \operatorname{tr} \mathbf{A}_1^2. \tag{4.83}$$

Hence, we have the result:

$$\nabla \cdot \mathbf{A}_1^2 = \nabla \left(\frac{1}{2} tr \; \mathbf{A}_1^2 \right). \tag{4.84}$$

To show that the divergence of $\mathbf{A}_2$ can be expressed as a gradient of a scalar, we need the results of Giesekus [23] and Pipkin [24]

$$\nabla \cdot \mathbf{A}_2 = \nabla \left(\frac{1}{\eta_0} \frac{dp_N}{dt} + \frac{3}{4} tr \; \mathbf{A}_1^2 \right). \tag{4.85}$$

Thus, a plane creeping flow in a Newtonian fluid is also a solution to the plane creeping flow problem in a second order fluid.

4.9 Green–Rivlin Expansion

There is an expansion proposed by Green and Rivlin [26] based on the integral of the strain history

$$\mathbf{G}(s) = \mathbf{C}_t(t-s) - \mathbf{I}, \quad 0 \le s < \infty, \tag{4.86}$$

which is regarded to be small in some sense (i.e., the relative strain $\mathbf{C}_t(t-s)$ is near to identity, the undeformed state). The expansion is quite unwieldy, consisting multiple integral terms. We record the first term here

$$\mathbf{S}(t) = \int_0^\infty \mu(s) \, \mathbf{G}(s) \, ds, \tag{4.87}$$

which is called the finite linear viscoelasticity. Here, $\mu(s)$ is the memory kernel, a decreasing function of s (distant past is less important – fluid has fading memory). Usually an exponential memory function is chosen:

$$\mu(s) = -\frac{G}{\lambda} e^{-s/\lambda}, \tag{4.88}$$

where G is a modulus and λ is a relaxation time. A multiple relaxation (exponential) mode is sometimes used as well.

Problems

Problem 4.1 In a simple shear deformation, the displacement field takes the form

$$v_1 = \gamma y, \quad v_2 = 0, \quad v_3 = 0, \tag{4.89}$$

where γ is the amount of shear. Find the elastic stress, in particular, the shear stress. This justifies calling μ the shear modulus.

Problem 4.2 In a uni-axial extension, the displacement field is given by

$$v_1 = \varepsilon x, \quad v_2 = -\nu\varepsilon y, \quad v_3 = -\nu\varepsilon z, \tag{4.90}$$

where ε is the elongational strain, and ν is the amount of lateral contraction due to the axial elongation, called Poisson's ratio. If the lateral stresses are zero, show that

$$\nu = \frac{\lambda}{2\left(\lambda + \mu\right)}, \quad \mu = \frac{E}{2\left(1 + \nu\right)}, \tag{4.91}$$

where E is the Young's modulus, i.e., $T_{xx} = E\varepsilon$.

Problem 4.3 Show that a material element $d\mathbf{X} = dX\mathbf{P}$, where $\mathbf{P}$ is a unit vector, is stretched according to

$$dx^2 = dX^2 \mathbf{C} : \mathbf{PP}.$$

Thus when $\mathbf{P}$ is randomly distributed in space, the average amount of stretch is

$$\langle dx^2 \rangle = \frac{1}{3} dX^2 \operatorname{tr} \mathbf{C},$$

and therefore $\frac{1}{3} \operatorname{tr} \mathbf{C}$ can be used as a definition of the Weissenberg number.

Problem 4.4 Let $\mathbf{f}$ be a vector valued, isotropic polynomial of a symmetric tensor $\mathbf{S}$ and a vector $\mathbf{v}$. Use the integrity basis in 4.36 to prove that

$$\mathbf{f}(\mathbf{S}, \mathbf{v}) - [f_0 \mathbf{1} + f_1 \mathbf{S} + f_2 \mathbf{S}^2]\mathbf{v},$$

where the scalar valued coefficients are polynomials in the six invariants involving only $\mathbf{S}$ and $\mathbf{v}$ in the list 4.36.

Problem 4.5 Consider a simple shear deformation, where

$$x = X + \gamma Y, \quad y = Y, \quad z = Z. \tag{4.92}$$

Show that the strain tensor $\mathbf{B}$ and its inverse are given by

$$[\mathbf{B}] = \begin{bmatrix} 1+\gamma^2 & \gamma & 0 \\ \gamma & 1 & 0 \\ 0 & 0 & 1 \end{bmatrix}, \quad [\mathbf{B}^{-1}] = \begin{bmatrix} 1 & -\gamma & 0 \\ -\gamma & 1+\gamma^2 & 0 \\ 0 & 0 & 1 \end{bmatrix}. \tag{4.93}$$

Consequently, show that the stress tensor is given by

$$[\mathbf{T}] = -P[\mathbf{I}] + \begin{bmatrix} \beta_1\left(1+\gamma^2\right) - \beta_2 & \left(\beta_1 + \beta_2\right)\gamma & 0 \\ \left(\beta_1 + \beta_2\right)\gamma & \beta_1 - \beta_2\left(1+\gamma^2\right) & 0 \\ 0 & 0 & \beta_1 - \beta_2 \end{bmatrix}, \tag{4.94}$$

where P is the hydrostatic pressure. Thus, show that the shear stress and the normal stress differences are

$$S = (\beta_1 + \beta_2)\gamma, \quad N_1 = (\beta_1 + \beta_2)\gamma^2, \quad N_2 = -\beta_2\gamma^2. \tag{4.95}$$

Deduce that the linear shear modulus of elasticity is

$$G = \lim_{\gamma \to 0} (\beta_1 + \beta_2). \tag{4.96}$$

The ratio

$$\frac{N_1}{S} = \gamma \tag{4.97}$$

is independent of the material properties. Such a relation is called *universal*.

Problem 4.6 In a uniaxial elongational deformation, where (in cylindrical coordinates)

$$R = \lambda^{1/2} r, \quad \Theta = \theta, \quad Z = \lambda^{-1} z, \tag{4.98}$$

show that the inverse deformation gradient is

$$\left[\mathbf{F}^{-1}\right] = \begin{bmatrix} \frac{\partial R}{\partial r} & \frac{1}{r}\frac{\partial R}{\partial \theta} & \frac{\partial R}{\partial z} \\ 0 & \frac{R}{r} & 0 \\ \frac{\partial Z}{\partial r} & \frac{1}{r}\frac{\partial Z}{\partial \theta} & \frac{\partial Z}{\partial z} \end{bmatrix} = \begin{bmatrix} \lambda^{1/2} & 0 & 0 \\ 0 & \lambda^{1/2} & 0 \\ 0 & 0 & \lambda^{-1} \end{bmatrix}. \tag{4.99}$$

Consequently, the strains are

$$\left[\mathbf{B}^{-1}\right] = \left[\mathbf{F}^{-T}\mathbf{F}^{-1}\right] = \begin{bmatrix} \lambda & 0 & 0 \\ 0 & \lambda & 0 \\ 0 & 0 & \lambda^{-2} \end{bmatrix}, \quad [\mathbf{B}] = \begin{bmatrix} \lambda^{-1} & 0 & 0 \\ 0 & \lambda^{-1} & 0 \\ 0 & 0 & \lambda^2 \end{bmatrix}. \tag{4.100}$$

Thus the total stress tensor is

$$[\mathbf{T}] = -P[\mathbf{I}] + \begin{bmatrix} \beta_1\lambda^{-1} - \beta_2\lambda & 0 & 0 \\ 0 & \beta_1\lambda^{-1} - \beta_2\lambda & 0 \\ 0 & 0 & \beta_1\lambda^2 - \beta_2\lambda^{-2} \end{bmatrix}. \tag{4.101}$$

Under the condition that the lateral tractions are zero, i.e., $T_{rr} = 0$, the pressure can be found, and thus show that the tensile stress is

$$T_{zz} = -P + \beta_1\lambda^2 - \beta_2\lambda^{-2} = \beta_1\lambda^2 - \beta_2\lambda^{-2} - \beta_1\lambda^{-1} + \beta_2\lambda$$
$$= \left(\lambda^2 - \lambda^{-1}\right)\left(\beta_1 + \beta_2\lambda^{-1}\right). \tag{4.102}$$

This tensile stress is the force per unit area in the deformed configuration. As $r = \lambda^{1/2}R$, the corresponding force per unit area in the undeformed configuration is

$$T_{ZZ} = T_{zz}\lambda^{-1} = \left(\lambda - \lambda^{-2}\right)\left(\beta_1 + \beta_2\lambda^{-1}\right). \tag{4.103}$$

Problem 4.7 Show that

$$\mathbf{C}(\tau) = \mathbf{F}(t)^T \mathbf{C}_t(\tau) \mathbf{F}(t).$$ (4.104)

Problem 4.8 Consider a simple shear flow

$$u = \dot{\gamma}y, \quad v = 0, \quad w = 0.$$ (4.105)

Show that the stress in the second-order model is given by

$$[\mathbf{S}] = \eta_0 \begin{bmatrix} 0 & \dot{\gamma} & 0 \\ \dot{\gamma} & 0 & 0 \\ 0 & 0 & 0 \end{bmatrix} + (\nu_1 + \nu_2) \begin{bmatrix} \dot{\gamma}^2 & 0 & 0 \\ 0 & \dot{\gamma}^2 & 0 \\ 0 & 0 & 0 \end{bmatrix} - \frac{\nu_1}{2} \begin{bmatrix} 0 & 0 & 0 \\ 0 & 2\dot{\gamma}^2 & 0 \\ 0 & 0 & 0 \end{bmatrix}.$$ (4.106)

Thus, the three viscometric functions are

$$S = \eta_0 \dot{\gamma}, \quad N_1 = S_{11} - S_{22} = \nu_1 \dot{\gamma}^2, \quad N_2 = S_{22} - S_{33} = \nu_2 \dot{\gamma}^2.$$ (4.107)

Problem 4.9 Consider a second-order fluid in an elongational flow

$$u = \dot{\varepsilon}x, \quad v = -\frac{\dot{\varepsilon}}{2}y, \quad w = -\frac{\dot{\varepsilon}}{2}z.$$ (4.108)

Show that the stress is given by

$$[\mathbf{S}] = \eta_0 \begin{bmatrix} 2\dot{\varepsilon} & 0 & 0 \\ 0 & -\dot{\varepsilon} & 0 \\ 0 & 0 & -\dot{\varepsilon} \end{bmatrix} + (\nu_1 + \nu_2) \begin{bmatrix} 4\dot{\varepsilon}^2 & 0 & 0 \\ 0 & \dot{\varepsilon}^2 & 0 \\ 0 & 0 & \dot{\varepsilon}^2 \end{bmatrix} - \frac{\nu_1}{2} \begin{bmatrix} 4\dot{\varepsilon}^2 & 0 & 0 \\ 0 & \dot{\varepsilon}^2 & 0 \\ 0 & 0 & \dot{\varepsilon}^2 \end{bmatrix}.$$

(4.109)

Consequently the elongational viscosity is given by

$$\eta_E = \frac{S_{xx} - S_{yy}}{\dot{\varepsilon}} = 3\eta_0 + 3\left(\frac{\nu_1}{2} + \nu_2\right)\dot{\varepsilon}.$$ (4.110)

Problem 4.10 Show that, for potential flows,

$$\nabla \cdot \mathbf{A}_1^2 = \frac{1}{2}\nabla(tr\ \mathbf{A}_1^2), \quad \nabla \cdot \mathbf{A}_2 = \frac{3}{4}\nabla(tr\ \mathbf{A}_1^2).$$ (4.111)

Problem 4.11 For steady two-dimensional incompressible flows, a stream function $\psi = \psi(x, y)$ can be defined such that the velocity components u and v can be expressed as

$$u = \frac{\partial \psi}{\partial y}, \quad v = -\frac{\partial \psi}{\partial x}.$$ (4.112)

Obtain the stresses in a second-order fluid in terms of ψ. Substitute the stresses into the equations of motion and eliminating the pressure term through the use of the equality of the mixed partial derivatives, i.e., $p_{,xy} = p_{,yx}$, to obtain

$$\eta_0 \triangle^2 \psi - \frac{1}{2}\Psi_1^0\ \mathbf{v} \cdot \nabla(\triangle^2 \psi) = 0,$$ (4.113)

where $\triangle^2$ is the two-dimensional biharmonic operator. Deduce that a Newtonian velocity field is also a velocity field for the second-order fluid. The result is due to Tanner [25].

Problem 4.12 In a simple shear flow,

$$u = \dot{\gamma}(t)\,y, \quad v = 0, \quad w = 0,$$

show that the path lines $\xi(\tau) = (\xi, \psi, \zeta)$ are given by

$$\xi(\tau) = x + y\gamma(t,\tau), \quad \psi(\tau) = y, \quad \zeta(\tau) = z, \tag{4.114}$$

where

$$\gamma(t,\tau) = \int_t^\tau \dot{\gamma}(s)\,ds. \tag{4.115}$$

Find the relative strain tensor, the stress tensor and show that the shear stress and the normal stress differences are given by

$$S_{12}(t) = \int_0^\infty \mu(s)\,\gamma(t, t - s)\,ds, \tag{4.116}$$

$$N_1(t) = \int_0^\infty \mu(s)\,[\gamma(t, t - s)]^2\,ds = -N_2. \tag{4.117}$$

Investigate the case where the shear rate is constant and sinusoidal in time for the memory function in (4.88).

5. Inelastic Models and Linear Viscoelasticity
Some practical engineering models

We have seen some of the classical constitutive equations introduced in the last three centuries, and explored some of the issues in the general formulation of constitutive equations in the last chapter. There, we mention that the general constitutive principles should be taken as guidelines only; they should emerge from the physics of the constitutive model. Only when the physics are so complex that we must guess which variables to put in our continuum model, and then these principles are applied to produce something useful. In engineering, the emphasis is to produce numbers for the design process. Therefore simple models with the "right" physics tend to be favoured. "Right" physics here means accounting for the flow process to be modelled. If the flow process does not call for certain phenomena, then those may be left out in the constitutive modelling. In addition, engineers do not have any qualms in using empirical data to supplement a constitutive equation, as long as the correct physical framework has already been incorporated. Remember, simplicity in the correct theoretical framework, with enough empirical inputs to ensure a quantitative prediction is for what we are striving.

5.1 Inelastic Fluids

When the flow phenomena are dominated by the viscosity alone, then it makes sense to model the viscosity function accurately. *Inelastic, or generalized Newtonian fluids* are those for which the extra stress tensor is proportional to the strain rate tensor, but the "constant" of proportionality (the viscosity) is allowed to depend on the strain rate:

$$\mathbf{S} = 2\eta\left(\dot{\gamma}\right)\mathbf{D}, \tag{5.1}$$

where $\dot{\gamma} = \sqrt{2\,\mathrm{tr}\,\mathbf{D}^2}$ is called the generalized strain rate. The inelastic model possesses neither memory nor elasticity, and therefore it is unsuitable for transient flows, or flows that call for elasticity effects. It is only useful in steady shear flows where an accurate representation of the viscosity is paramount. Depending on the functional form of $\eta\left(\dot{\gamma}\right)$, one can get different non-Newtonian behaviours, see Fig. 5.1.

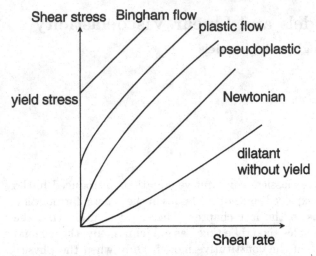

Fig. 5.1. Different non-Newtonian behaviours

Bingham fluids are those that can support a yield stress. When the shear stress exceeds this yield value, the fluid flows like a Newtonian fluid, with constant viscosity. Plastic fluids are yield stress type of fluid, with flow viscosity decreasing with shear rate (shear thinning). The term "pseudoplastic" means that the viscosity decreases with shear rates (shear thinning). The opposite of pseudoplastic is dilatant (shear thickening).

5.1.1 Carreau Model

Different forms for the viscosity have been proposed for pseudoplastic fluids, the most popular one is the Carreau model:

$$\eta\left(\dot{\gamma}\right) = \eta_\infty + \frac{\eta_0 - \eta_\infty}{\left(1 + \Gamma^2 \dot{\gamma}^2\right)^{(1-n)/2}}. \tag{5.2}$$

There are four parameters: η_∞ is the infinite-shear-rate viscosity, η_0 is the zero-shear-rate viscosity, n is called the power-law index ($n - 1$ is the slope of $(\eta - \eta_\infty)/(\eta_0 - \eta_\infty)$ versus $\dot{\gamma}$ in log-log plot), and Γ is a time constant. A graph of (5.2) is illustrated in Fig. 5.2.

The model is meant for shear-thinning fluids and therefore $0 < n \leq 1$. $n = 1$ represents the Newtonian behaviour.

5.1.2 Power-law Model

Included in the Carreau model is the power-law model:

$$\eta\left(\dot{\gamma}\right) = k\left|\dot{\gamma}\right|^{n-1}, \tag{5.3}$$

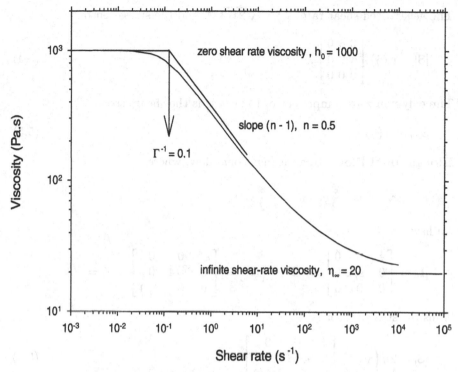

Fig. 5.2. A typical plot of the Carreau viscosity

where k is called the consistency and n the power-law index. When $\eta_\infty = 0$ and at high shear rate, the Carreau model (5.2) reduces to the power-law model with power-law index n and consistency $k = \eta_0 \Gamma^{n-1}$. The power-law model breaks down in regions where the shear rate is zero - in these regions the stress is unbounded.

Simple Shear Flow. In a simple shearing flow, where the fluid is confined between two plates, and the top plate is moving with a velocity U, the velocity field takes the form

$$u = \dot{\gamma}y, \quad v = 0, \quad w = 0, \quad \dot{\gamma} = U/h$$

the velocity gradient and the strain rate tensors are

$$[\mathbf{L}] = [\nabla \mathbf{u}^T] = \begin{bmatrix} 0 & \dot{\gamma} & 0 \\ 0 & 0 & 0 \\ 0 & 0 & 0 \end{bmatrix}, \quad [\mathbf{D}] = \frac{1}{2}(\mathbf{L} + \mathbf{L}^T) = \frac{1}{2}\begin{bmatrix} 0 & \dot{\gamma} & 0 \\ \dot{\gamma} & 0 & 0 \\ 0 & 0 & 0 \end{bmatrix}.$$

and

$$[\mathbf{D}^2] = \frac{1}{4}\begin{bmatrix} 0 & \dot{\gamma} & 0 \\ \dot{\gamma} & 0 & 0 \\ 0 & 0 & 0 \end{bmatrix}\begin{bmatrix} 0 & \dot{\gamma} & 0 \\ \dot{\gamma} & 0 & 0 \\ 0 & 0 & 0 \end{bmatrix} = \frac{1}{4}\begin{bmatrix} \dot{\gamma}^2 & 0 & 0 \\ 0 & \dot{\gamma}^2 & 0 \\ 0 & 0 & 0 \end{bmatrix}.$$

The generalized shear rate is $\dot\gamma = \sqrt{2\,\mathrm{tr}\,\mathbf{D}^2}$, and the stress tensor is

$$[\mathbf{S}] = \eta\,(\dot\gamma)\begin{bmatrix} 0 & \dot\gamma & 0 \\ \dot\gamma & 0 & 0 \\ 0 & 0 & 0 \end{bmatrix}. \tag{5.4}$$

The only non-zero component of the stress is the shear stress,

$$S_{xy} = \eta\,(\dot\gamma)\,\dot\gamma. \tag{5.5}$$

Elongational Flow. In an elongational flow, where

$$u = \dot\varepsilon x, \quad v = -\frac{\dot\varepsilon}{2}y, \quad w = -\frac{\dot\varepsilon}{2}z,$$

we have

$$[\mathbf{L}] = \begin{bmatrix} \dot\varepsilon & 0 & 0 \\ 0 & -\dot\varepsilon/2 & 0 \\ 0 & 0 & 0 \end{bmatrix} = [\mathbf{D}], \quad [\mathbf{D}^2] = \begin{bmatrix} \dot\varepsilon^2 & 0 & 0 \\ 0 & \dot\varepsilon^2/4 & 0 \\ 0 & 0 & \dot\varepsilon^2/4 \end{bmatrix}, \quad \dot\gamma = \sqrt{3}\dot\varepsilon.$$

Thus the stress tensor is

$$[\mathbf{S}] = 2\eta\left(\sqrt{3}\dot\varepsilon\right)\begin{bmatrix} \dot\varepsilon & 0 & 0 \\ 0 & -\dot\varepsilon/2 & 0 \\ 0 & 0 & -\dot\varepsilon/2 \end{bmatrix} \tag{5.6}$$

and the elongational viscosity is given by

$$\eta_E = \frac{N_1}{\dot\varepsilon} = 3\eta\left(\sqrt{3}\dot\varepsilon\right). \tag{5.7}$$

It has been known that the elongational viscosity and the shear viscosity have dissimilar shape. Therefore, in-elastic models may not be suitable in processes where there are a mixture of shear and elongational flow components.

5.2 Linear Viscoelasticity

The concept of linear viscoelastic was originated with Maxwell (Fig. 5.3), who proposed the equation in 1867–68,

$$\frac{d\sigma}{dt} = E\frac{d\varepsilon}{dt} - \frac{\sigma}{\lambda}, \tag{5.8}$$

where σ is the (one-dimensional) stress, ε is the (one-dimensional) strain, E is the modulus of elasticity and λ is a time constant. When the relaxation time is zero, keeping the product $\eta = \lambda E$ constant, the Newtonian model is

Fig. 5.3. J.C. Maxwell (1844–1906) published his first scientific paper when he was thirteen. He set up Cavendish Laboratory at Cambridge in 1874, and died of cancer at an early age of 48. He united electricity and magnetism into the concept of electromagnetism. He also introduced the concept of stress relaxation in the kinetic theory of gases

recovered. And when the relaxation is infinitely large, a further integration yields the Hookean model.

Years later, Meyer (1874) introduced the equation

$$\sigma = G\gamma + \eta \frac{d\gamma}{dt}, \tag{5.9}$$

combining the solid response (G is the elastic modulus, γ is the shear strain) and liquid response (η is the viscosity) in one equation. This equation is now known as Kelvin–Voigt body – it should be called Kelvin–Meyer–Voigt (the full account can be found in Tanner and Walter [13]).

Fig. 5.4. Ludwig Boltzmann (1844–1906) was most famous for his atomic viewpoint and his invention of statistical mechanics. He is said never to have failed any student taking his course

Boltzmann criticized the lack of generality in Maxwell's and Meyer's work and proposed that the stress at the current time depends not only on the current strain, but on the past strains as well. It was assumed that a strain at a distant past contributes less to the stress than a more recent strain. This is recognized as the familiar concept of fading memory. Furthermore, linear superposition was assumed: supposing that the strain between times t' and $t' + dt'$, say $d\gamma(t')$, contributes $G(t - t') d\gamma(t')$ to the stress, then the total stress at time t is

$$\sigma(t) = \int_{-\infty}^{t} G(t - t') \, d\gamma(t') = \int_{-\infty}^{t} G(t - t') \, \dot{\gamma}(t') \, dt'. \tag{5.10}$$

Here, $G(t)$ is a decreasing function of time, the *relaxation modulus*, and $\dot{\gamma}$ is the shear rate. The three-dimensional version of this is

$$\mathbf{S}(t) = 2 \int_{-\infty}^{t} G(t - t') \mathbf{D}(t') \, dt', \tag{5.11}$$

$$S_{ij}(t) = 2 \int_{-\infty}^{t} G(t - t') D_{ij}(t') \, dt'.$$

The Newtonian liquid is recovered with the delta memory function

$$G(t - t') = \eta_0 \delta(t - t')$$

$$\therefore \quad \mathbf{S}(t) = 2 \int_{-\infty}^{t} \eta_0 \delta(t - t') \mathbf{D}(t') \, dt' = 2 \eta_0 \mathbf{D}(t). \tag{5.12}$$

The most-often used relaxation modulus function is the *Maxwell discrete relaxation spectrum*:

$$G(t) = \sum_{j=1}^{N} G_j e^{-t/\lambda_j}, \tag{5.13}$$

which consists of a discrete spectrum of relaxation times $\{G_j, \lambda_j\}$. The linear viscoelastic relation (5.11) is not objective; it is only valid at small strains.

5.2.1 Simple Shear Flow

In a flow with constant strain rate, the stress is

$$\mathbf{S}(t) = 2\eta(t)\mathbf{D}, \quad \eta(t) = \int_{-\infty}^{t} G(t - t') \, dt'. \tag{5.14}$$

With the Maxwell relaxation modulus (5.13),

$$\mathbf{S}(t) = \sum_{j=1}^{N} 2\eta_j \left(1 - e^{-t/\lambda_j}\right) \mathbf{D}, \quad \eta_j = G_j \lambda_j. \tag{5.15}$$

Now consider an oscillatory shear flow, where the top plate is sinusoidally displaced by a small amount $\delta \sin \omega t$. The top plate velocity is

$$U(t) = \delta \omega \cos \omega t.$$

The shear rate and the shear strain are

$$\dot{\gamma}(t) = \frac{\delta}{h} \omega \cos \omega t = \dot{\gamma}_0 \cos \omega t,$$

$$\gamma(t) = \frac{\delta}{h} \sin \omega t = \gamma_0 \sin \omega t, \quad \gamma_0 = \frac{\delta}{h} \ll 1, \quad \dot{\gamma}_0 = \omega \gamma_0. \tag{5.16}$$

The only non-zero component of the stress is the shear stress,

$$S_{12} = \int_{-\infty}^{t} G\left(t - t'\right) \dot{\gamma}_0 \cos \omega t' dt',$$

$$= \int_{0}^{\infty} \dot{\gamma}_0 G\left(s\right) \cos \omega \left(t - s\right) ds,$$

$$= \int_{0}^{\infty} \dot{\gamma}_0 G\left(s\right) \left[\cos \omega t \cos \omega s + \sin \omega t \sin \omega s\right] ds,$$

$$= G'\left(\omega\right) \gamma_0 \sin \omega t + \eta'\left(\omega\right) \dot{\gamma}_0 \cos \omega t,$$

where the coefficient in the strain, $G'\left(\omega\right)$, is called the *storage modulus*, and that in the strain rate, $\eta'\left(\omega\right)$, the *dynamic viscosity*. These are material functions of the frequency and are defined by

$$G'\left(\omega\right) = \int_{0}^{\infty} \omega G\left(s\right) \sin \omega s ds, \quad \eta'\left(\omega\right) = \int_{0}^{\infty} G\left(s\right) \cos \omega s ds. \tag{5.17}$$

The other two related quantities, the loss modulus, and the storage viscosity are defined as

$$G''\left(\omega\right) = \omega \eta'\left(\omega\right), \quad \eta''\left(\omega\right) = \frac{G'\left(\omega\right)}{\omega}. \tag{5.18}$$

Sometimes it is more convenient to work with complex numbers, and the complex modulus and complex viscosity are thus defined as

$$G^*\left(\omega\right) = G'\left(\omega\right) + iG''\left(\omega\right), \quad \eta^*\left(\omega\right) = \eta'\left(\omega\right) - i\eta''\left(\omega\right). \tag{5.19}$$

One can write

$$\gamma\left(t\right) = \gamma_0 e^{i\omega t}, \quad S_{12}\left(t\right) = G^* \gamma_0 e^{i\omega t}. \tag{5.20}$$

It should be remembered that the linearity is valid only at small strains, that is, both G' and η' should be independent of the strain amplitude γ_0; this ought to be tested before a frequency sweep is done. The strain amplitude for which linearity holds could be as large as 20% for polymer melts and solutions, and as small as 0.1% for biological materials, such as bread dough.

Figure 5.5 is a plot of the storage modulus and dynamic viscosity of a water-dough as functions of the frequency, where the strain amplitude was kept at 0.1%.

From the inverse Fourier transform of (5.17),

$$G\left(s\right) = \frac{2}{\pi} \int_{0}^{\infty} \eta'\left(\omega\right) \cos \omega s d\omega = \frac{2}{\pi} \int_{0}^{\infty} \frac{G'\left(\omega\right)}{\omega} \sin \omega s d\omega. \tag{5.21}$$

Inverting the dynamic properties for the relaxation modulus may be an ill-conditioned problem.

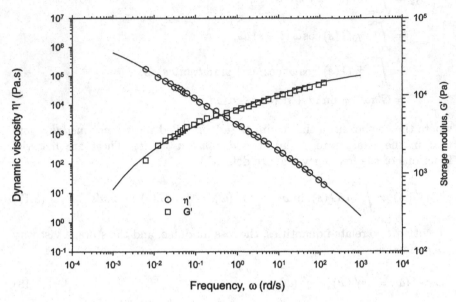

Fig. 5.5. Dynamic properties of flour-water dough. The strain amplitude was set at 0.1%

5.2.2 Step Strain

Consider the stress relaxation experiment after a step strain, Fig. 5.6. Suppose a shear strain of magnitude γ is imposed at time t_0 within a short period Δ, as sketched in Fig. 5.6. During this period the strain rate can be assumed constant, given by γ/Δ. The stress is

$$S_{12}(t) = \int_{t_0}^{t_0+\Delta} G(t-s)\frac{\gamma}{\Delta}ds. \tag{5.22}$$

Recall the mean-value theorem,

$$\int_{t_0}^{t_0+\Delta} f(s)\,ds = \Delta f(t_0+\zeta), \quad 0 \le \zeta \le \Delta.$$

When applied to (5.22), this yields

$$S_{12}(t) = \gamma G(t-t_0-\zeta).$$

When $\Delta \to 0$, $\zeta \to 0$, and the stress relaxes as does the modulus of relaxation (hence the name). In particular, when $t_0 = 0$,

$$S_{12}(t) = \gamma G(t). \tag{5.23}$$

The relaxation modulus is a material function and can be measured routinely on rheometer.

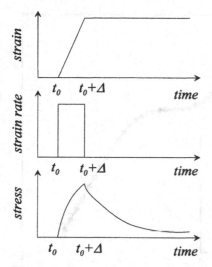

Fig. 5.6. Stress relaxation after a step strain

5.2.3 Relaxation Spectrum

We have met the discrete Maxwell relaxation spectrum, (5.13). The continuous version of this is the *relaxation spectrum*, defined by

$$G(t) = \int_0^\infty H(\lambda) e^{-t/\lambda} \frac{d\lambda}{\lambda} = \int_{-\infty}^\infty H(\lambda) e^{-t/\lambda} d\ln\lambda. \qquad (5.24)$$

In this, the relaxation time λ is supposedly evenly distributed on a logarithmic scale. The quantity $H(\lambda)$ is called the relaxation spectrum. Note that

$$
\begin{aligned}
G'(\omega) &= \int_0^\infty \omega \sin\omega s \int_0^\infty H(\lambda) e^{-t/\lambda} \frac{d\lambda}{\lambda} ds \\
&= \int_0^\infty \omega H(\lambda) \frac{d\lambda}{\lambda} \int_0^\infty e^{-t/\lambda} \sin\omega s \, ds \qquad (5.25)
\end{aligned}
$$

The last integral can be evaluated, giving

$$G'(\omega) = \int_{-\infty}^\infty \frac{\omega^2 \lambda^2}{1+\omega^2\lambda^2} H(\lambda) d\ln\lambda. \qquad (5.26)$$

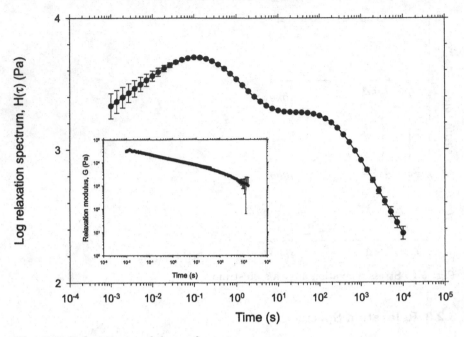

Fig. 5.7. Relaxation modulus and spectrum

Similarly,

$$\eta'(\omega) = \int_0^\infty \frac{H(\lambda)}{1+\omega^2\lambda^2} d\lambda. \tag{5.27}$$

Inversion of the data to find H is an ill-conditioned problem. In Fig. 5.7 the relaxation modulus, and its spectrum are shown for a dough-water flour, using a regularisation method of Weese [27].

5.3 Correspondence Principle

5.3.1 Quasi-Static Approximation

In the boundary-value problem for a linear viscoelastic fluid, the quasi-static approximation is used. Here, one ignores the inertia terms $\rho\ddot{\mathbf{u}}$. This is possible if the characteristic frequency is not too high (De $\ll$ 1). Let suppose the flow starts from time zero, before which the stress is zero. The equations of motion are

$$\mathbf{T}(t) = -P\mathbf{I} + \mathbf{S} = -P\mathbf{I} + \int_0^t G(t-t')\left(\nabla\mathbf{u}(t') + \nabla\mathbf{u}(t')^T\right)dt', \tag{5.28}$$

$$\nabla\cdot\mathbf{T} = -\nabla P + \nabla\cdot\mathbf{S} = 0, \quad \nabla\cdot\mathbf{u} = 0, \quad \mathbf{x}\in V \tag{5.29}$$

subjected to the boundary condition on the bounding surface S:

$$\mathbf{u}(\mathbf{x},t) = \mathbf{u}_0(t), \quad \mathbf{x} \in S. \tag{5.30}$$

We can take the Laplace transform of all the above equations to arrive at

$$\bar{\mathbf{T}}(s) = -\bar{P}\mathbf{I} + \bar{G}\left(\nabla\bar{\mathbf{u}} + \nabla\bar{\mathbf{u}}^T\right), \tag{5.31}$$

$$-\nabla\bar{P} + \bar{G}\nabla^2\bar{\mathbf{u}} = 0, \quad \nabla\cdot\bar{\mathbf{u}} = 0, \tag{5.32}$$

$$\bar{\mathbf{u}}(\mathbf{x},s) = \bar{\mathbf{u}}_0(s), \quad \mathbf{x} \in S, \tag{5.33}$$

where the overbar denotes a Laplace transform variable. i.e.,

$$\bar{\phi}(s) = \int_0^\infty e^{-st}\phi(t)\,dt. \tag{5.34}$$

Equations (5.31)–(5.33) are identical to those of the Newtonian (Stokes) problem, except that the viscosity is now replaced by $\bar{G}$. Thus the solution in the Laplace transform domain matches the Stokes solution. This is the essence of the *Correspondence Principle*. The solution in the physical domain is then obtained by inverting the Laplace transform. We give an example for the start-up of a circular Couette flow.

5.3.2 Circular Couette Flow

The velocity field is, in cylindrical coordinates (Fig. 5.8),

$$\mathbf{u} = \{0, r\Omega(r,t), 0\}, \tag{5.35}$$

where

$$\Omega(R_i,t) = \Omega_i, \quad \Omega(R_o,t) = 0. \tag{5.36}$$

The strain rate tensor is

$$[\mathbf{D}] = \frac{1}{2}\begin{bmatrix} 0 & r\dfrac{\partial\Omega}{\partial r} & 0 \\ r\dfrac{\partial\Omega}{\partial r} & 0 & 0 \\ 0 & 0 & 0 \end{bmatrix}.$$

The only non-zero component of the stress is the shear component

$$S_{r\theta} = \int_0^t G(t-t')\,r\frac{\partial\Omega}{\partial r}(r,t')\,dt'.$$

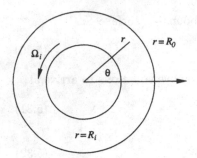

Fig. 5.8. Circular Couette flow

The balance of linear momentum requires

$$\frac{\partial}{\partial r}\left(r^2 S_{r\theta}\right) = 0, \quad \therefore \quad S_{r\theta} = \frac{M(t)}{2\pi r^2},$$

where $M(t)$ is a "constant" of integration. The torque is

$$\Gamma = \int_{R_i}^{R_o} 2\pi r^2 S_{r\theta} dr = M\left(R_o - R_i\right). \tag{5.37}$$

Taking the Laplace transform,

$$\bar{S}_{r\theta} = \bar{G}r\frac{\partial\bar{\Omega}}{\partial r} \quad \therefore \quad \frac{\partial\bar{\Omega}}{\partial r} = \frac{\bar{M}}{2\pi\bar{G}r^3}.$$

Integrating

$$\bar{\Omega} = C - \frac{\bar{M}}{4\pi r^2\bar{G}},$$

where C is an integration constant. Applying the boundary conditions (5.36),

$$\bar{\Omega} = \frac{\bar{M}}{4\pi R_o^2\bar{G}}\left(1 - \frac{R_o^2}{r^2}\right), \quad \bar{\Omega}_i = \frac{\bar{M}}{4\pi R_o^2\bar{G}}\left(1 - \frac{R_o^2}{R_i^2}\right). \tag{5.38}$$

Note that

$$\frac{\bar{\Omega}}{\bar{\Omega}_i} = \frac{1 - R_o^2/r^2}{1 - R_o^2/R_i^2}, \tag{5.39}$$

which is the Stokes solution. In the Newtonian case, $G(t) = \eta_0\delta(t)$, giving $\bar{G} = \eta_0$. Hence,

$$M_N = \frac{4\pi R_o^2\eta_0\Omega_i}{1 - R_o^2/R_i^2}. \tag{5.40}$$

For the linear viscoelastic case,

$$\bar{M} = \frac{4\pi R_o^2}{1 - R_o^2/R_i^2}\bar{G}\bar{\Omega}_i,$$

$$M(t) = \frac{4\pi R_o^2}{1 - R_o^2/R_i^2}\int_0^t G(t - t')\,\Omega_i(t')\,dt'. \tag{5.41}$$

5.4 Mechanical Analogs

In the older rheology literature, one finds mechanical analogs for linear vis-coelastic behaviours, springs for solid behaviour and dashpots for viscous behaviour. We illustrate this with a few popular models.

The corresponding multimode Kelvin–Voigt–Meyer model (5.48) is written as

$$\gamma_{ij} = \sum_{n=1}^{N} \gamma_{ij}^{(n)}, \quad \gamma_{ij}^{(n)} + \lambda_n \dot{\gamma}_{ij}^{(n)} = \frac{S_{ij}}{G_n}. \tag{5.42}$$

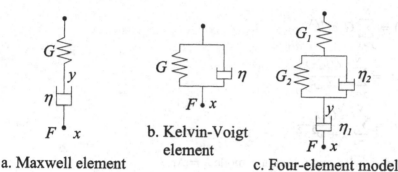

a. Maxwell element

b. Kelvin-Voigt element

c. Four-element model

Fig.5.9a–c. Mechanical analogs of linear viscoelastic behaviours

Problems

Problem 5.1 Show that, with the relaxation modulus function (5.13), the relation (5.11) is equivalent to

$$\mathbf{S} = \sum_{j=1}^{N} \mathbf{S}^{(j)}, \tag{5.43}$$

$$\mathbf{S}^{(j)} + \lambda_j \dot{\mathbf{S}}^{(j)} = 2\eta_j \mathbf{D}, \quad \eta_j = G_j \lambda_j.$$

This relation is called the linear Maxwell equation. Equation (5.43) is equivalent to (5.9).

Problem 5.2 Show that the shear stress can be expressed as

$$S_{12} = |G^*| \sin(\omega t + \phi), \quad \tan\phi = \frac{G''}{G'}, \tag{5.44}$$

where $\tan\phi$ is called the loss tangent.

Problem 5.3 Verify that for the spectrum

$$H(\lambda) = \cos^2(n\lambda),$$

$$\eta'(\omega) = \int_0^\infty \frac{H(\omega)}{1 + \lambda^2\omega^2} d\omega = \frac{\pi}{4\omega}\left(1 - e^{-2n/\omega}\right).$$

At large n, the data η' is smooth, but the spectrum is highly oscillatory. Conclude that the inverse problem of finding $H(\lambda)$, given the data η' in the chosen form is ill-conditioned – that is, a small variation in the data (in the exponentially small term) may lead to a large variation in the solution.

Problem 5.4 For the Maxwell discrete relaxation spectrum (5.13), show that

$$G(t) = \sum_{j=1}^N G_j e^{-t/\lambda_j},$$

$$G'(\omega) = \sum_{j=1}^N \frac{G_j\omega^2\lambda_j^2}{1 + \omega^2\lambda_j^2},$$

$$\eta'(\omega) = \sum_{j=1}^N \frac{G_j\lambda_j}{1 + \omega^2\lambda_j^2}. \tag{5.45}$$

In particular, with one relaxation mode $\lambda = \lambda_1$,

$$\tan\phi = \frac{1}{\omega\lambda} \tag{5.46}$$

deduce that as $\omega = 0 \to \infty$, the response goes from fluid ($\phi = 0$) to solid behaviour ($\phi = \pi/2$).

Problem 5.5 Suppose we have a Maxwell material with one relaxation time,

$$G(t) = \frac{\eta_0}{\lambda} e^{-t/\lambda}.$$

and $\Omega_i = $ constant. Show that

$$\frac{M(t)}{M_N} = 1 - e^{-t/\lambda}. \tag{5.47}$$

Problem 5.6 In the mechanical model depicted in Fig. 5.9a, show that the force F is given by

$$F = \eta\dot{y} = G(x - y).$$

By eliminating y, show that

$$F + \frac{\eta}{G}\dot{F} = \eta\dot{x}.$$

Thus, if F is identified with the stress and x, the strain, then one obtains the Maxwell model.

Problem 5.7 In the mechanical model of Fig. 5.9b, show that

$$Gx + \eta\dot{x} = F.$$

Identify x with the strain, and F with the stress, one obtains the Kelvin–Voigt–Meyer model:

$$G\gamma_{ij} + \eta\dot{\gamma}_{ij} = S_{ij}. \tag{5.48}$$

The Kelvin–Voigt–Meyer material is a solid (it can support a shear stress indefinitely without deforming).

Problem 5.8 Show that the mechanical analog of Fig. 5.9c leads to the following four-element model

$$S_{ij} + a_1\dot{S}_{ij} + a_2\ddot{S}_{ij} = b_0\gamma_{ij} + b_1\dot{\gamma}_{ij} + b_2\ddot{\gamma}_{ij}, \tag{5.49}$$

where

$$a_1 = \frac{\eta_1}{G_2}\left(1 + \frac{G_2}{G_1} + \frac{\eta_2}{\eta_1}\right), \quad a_2 = \frac{\eta_1\eta_2}{G_1G_2}, \quad b_0 = 0, \quad b_1 = \eta_1, \quad b_2 = \frac{\eta_1\eta_2}{G_2}.$$

6. Steady Viscometric Flows

Shear flows

There is a class of flows for which the kinematics and the stress can be determined for the simple fluid. These flows are equivalent to the simple shearing flow. Ericksen [28] called them *laminar shear flows*, but the current terms used to describe these flows are *viscometric flows* [29]. Here, we review this class of flows.

6.1 Kinematics

Consider a simple shear flow with the kinematics

$$u = \dot{\gamma} y, \quad v = 0, \quad w = 0, \tag{6.1}$$

where the shear rate $\dot{\gamma}$ is a constant. This flow has the velocity gradient tensor

$$[\mathbf{L}] = \begin{bmatrix} 0 & \dot{\gamma} & 0 \\ 0 & 0 & 0 \\ 0 & 0 & 0 \end{bmatrix},$$

which obeys $\mathbf{L}^2 = \mathbf{0}$. For this flow, the only non-trivial Rivlin–Ericksen tensors are $\mathbf{A}_1$ and $\mathbf{A}_2$; the rest of the Rivlin–Ericksen tensors are nil,

$$[\mathbf{A}_1] = \begin{bmatrix} 0 & \dot{\gamma} & 0 \\ \dot{\gamma} & 0 & 0 \\ 0 & 0 & 0 \end{bmatrix}, \quad [\mathbf{A}_2] = \begin{bmatrix} 0 & 0 & 0 \\ 0 & 2\dot{\gamma}^2 & 0 \\ 0 & 0 & 0 \end{bmatrix}.$$

Consequently, the relative right Cauchy–Green tensor is quadratic in the time lapse:

$$\mathbf{C}_t (t - s) = \mathbf{I} - s\mathbf{A}_1 + \frac{s^2}{2}\mathbf{A}_2, \quad 0 \leq s. \tag{6.2}$$

Ericksen [28] referred to flows obeying (6.2) laminar shear flows, Coleman [29] called them viscometric flows. Yin and Pipkin [30] embarked on a search for all such flows and now our knowledge of them is essentially complete. One can write for (6.1),

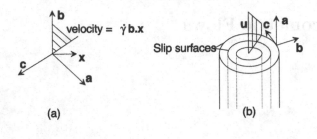

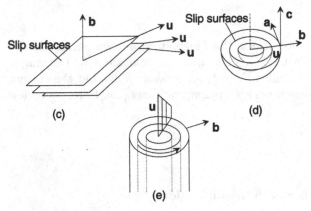

Fig. 6.1. Viscometric flows

$$\mathbf{u} = \dot{\gamma}\,(\mathbf{b} \cdot \mathbf{x})\,\mathbf{a}. \tag{6.3}$$

In the simple shear flow considered here, $\mathbf{a} = \mathbf{e}_1$, $\mathbf{b} = \mathbf{e}_2$. We now define viscometric flows are those where the velocity field obeys (6.3), for three mutually orthogonal unit vectors $\mathbf{a}$, $\mathbf{b}$, and $\mathbf{c}$, refer to Fig. 6.1.

The three directions $\mathbf{a}$, $\mathbf{b}$, and $\mathbf{c}$ are called the shear axes; $\mathbf{a}$ is the direction of shear, $\mathbf{b}$ is the direction of shear rate, and $\mathbf{c}$ is the vorticity axis. The motion can be visualized as the relative sliding motion of a stack of playing cards, each card represents a slip surface $\mathbf{b} \cdot \mathbf{x} =$ constant.

In Fig. 6.1, all such flows are sketched: (a) simple shearing flow, (b) steady parallel flow, (c) rectilinear flow, (d) circular flow, and (d) helical flow. Shear flow has already been considered, we now briefly look at the rest.

6.1.1 Steady Parallel Flow

In this flow (Fig. 6.1b), the velocity takes the form

$$\mathbf{u} = w\,(x, y)\,\mathbf{k}, \tag{6.4}$$

where $\mathbf{a} = \mathbf{k}$ is a unit vector in the z-direction, the slip surfaces are cylinders with constant $w\,(x, y)$. The velocity gradient is

$$(\nabla \mathbf{u})^T = \frac{\partial w}{\partial x}\mathbf{ki} + \frac{\partial w}{\partial y}\mathbf{kj} = \dot{\gamma}\mathbf{ab}, \tag{6.5}$$

with

$$\dot{\gamma}\mathbf{b} = \frac{\partial w}{\partial x}\mathbf{i} + \frac{\partial w}{\partial y}\mathbf{j}, \quad \dot{\gamma}^2 = \left(\frac{\partial w}{\partial x}\right)^2 + \left(\frac{\partial w}{\partial y}\right)^2. \tag{6.6}$$

The material derivative of $\dot{\gamma}$ is zero, i.e., the shear rate is constant along each streamline.

6.1.2 Rectilinear flow

Another class of rectilinear flows is that in which the velocity field takes the form (Fig. 6.1c)

$$\mathbf{u} = u(z)\mathbf{i} + v(z)\mathbf{j}. \tag{6.7}$$

These flows have parallel plane surfaces $z =$ constant, like a pack of playing cards. The velocity gradient is

$$(\nabla \mathbf{u})^T = \frac{\partial u}{\partial z}\mathbf{ik} + \frac{\partial v}{\partial z}\mathbf{jk} = \dot{\gamma}\mathbf{ab}, \tag{6.8}$$

where the direction of the shear rate is $\mathbf{b} = \mathbf{k}$, and

$$\dot{\gamma}\mathbf{a} = \frac{\partial u}{\partial z}\mathbf{i} + \frac{\partial v}{\partial z}\mathbf{j}, \quad \dot{\gamma}^2 = \left(\frac{\partial u}{\partial z}\right)^2 + \left(\frac{\partial v}{\partial z}\right)^2. \tag{6.9}$$

Again the shear rate remains constant along each streamline.

6.1.3 Axial Fanned Flow

In the axial fanned flow, the velocity takes the form

$$\mathbf{u} = c\theta\mathbf{k}, \quad 0 \le \theta \le 2\pi, \quad \dot{\gamma} = c/r, \tag{6.10}$$

where c is a constant and $\theta = \tan^{-1}(y/x)$.

6.1.4 Helical Flow

In the axial translation, rotation and screw motions of coaxial circular slip surfaces (Fig. 6.1e), the velocity takes the form

$$\mathbf{u} = r\omega(r)\mathbf{e}_\theta + u(r)\mathbf{e}_z. \tag{6.11}$$

The velocity gradient is

$$(\nabla \mathbf{u})^T = r \frac{\partial \omega}{\partial r} \mathbf{e}_\theta \mathbf{e}_r + \frac{\partial u}{\partial r} \mathbf{e}_z \mathbf{e}_r = \dot{\gamma} \mathbf{a} \mathbf{b}, \tag{6.12}$$

where $\mathbf{b} = \mathbf{e}_r$, and

$$\dot{\gamma} \mathbf{a} = r \frac{\partial \omega}{\partial r} \mathbf{e}_\theta + \frac{\partial u}{\partial r} \mathbf{e}_z, \quad \dot{\gamma}^2 = r^2 \left(\frac{\partial \omega}{\partial r} \right)^2 + \left(\frac{\partial u}{\partial r} \right)^2. \tag{6.13}$$

The shear rate remains constant along each streamline. These flows include

1. Circular pipe flow, or Poiseuille flow, when the flow occurs in a circular pipe, or annular flow, when it occurs between two concentric cylinders. Here $\omega = 0$.
2. Circular Couette flow, when the flow occurs between concentric cylinders, one or both rotating, and $u = 0$.
3. Helical flow, when both rotational and translational components are present. Since the angular velocity is ω, a particle covers 2π radians in $2\pi/\omega$ seconds, while rising $2\pi u/\omega$. This rise is constant on each cylinder surface.

6.1.5 Helicoidal Flow

If all the helical paths in the previous flow have the same rise per turn, then the slip surfaces need not be helical but can be general helicoids. The velocity field takes the form

$$\mathbf{u} = (r\mathbf{e}_\theta + c\mathbf{e}_z)\, \omega\, (r, z - c\theta), \tag{6.14}$$

where c is a constant and ω is a function of r and $z - c\theta$. All the helices have the same rise per turn, i.e., $2\pi/c$. The velocity gradient is

$$\begin{aligned}
(\nabla \mathbf{u})^T &= [\nabla (r\omega \mathbf{e}_\theta + c\omega \mathbf{e}_z)]^T = (\nabla \omega\, (r\mathbf{e}_\theta + c\mathbf{e}_z) + \omega\, (\mathbf{e}_r \mathbf{e}_\theta - \mathbf{e}_\theta \mathbf{e}_r))^T \\
&= (r\mathbf{e}_\theta + c\mathbf{e}_z)\, \nabla \omega + \omega\, (\mathbf{e}_\theta \mathbf{e}_r - \mathbf{e}_r \mathbf{e}_\theta) \\
&= \dot{\gamma} \mathbf{a} \mathbf{b}.
\end{aligned} \tag{6.15}$$

6.2 Stresses in Steady Viscometric Flows

It is remarkable that the stresses in steady viscometric flows can be determined completely for isotropic simple fluids. Take, for example, the simple shear flow where $\mathbf{u} = \dot{\gamma} y \mathbf{i}$. The general form of the stress tensor is

$$[\mathbf{S}] = \begin{bmatrix} S_{xx} & S_{xy} & 0 \\ S_{xy} & S_{yy} & 0 \\ 0 & 0 & S_{zz} \end{bmatrix}.$$

The shear stress S_{xy} must be an odd function of the shear rate based on physical ground alone.[1] Thus we can write

$$S_{xy} = \dot{\gamma}\eta\left(\dot{\gamma}\right), \tag{6.16}$$

where the viscosity, defined as $\eta = S_{xy}/\dot{\gamma}$, must be an even function of the shear rate:

$$\eta\left(-\dot{\gamma}\right) = \eta\left(\dot{\gamma}\right). \tag{6.17}$$

Reversing the direction of shear will not change the normal stress components, and therefore these will be even functions of the shear rate. The arbitrary pressure can be eliminated by taking the differences between these normal stresses. We thus define the first and the second normal stress differences by

$$S_{xx} - S_{yy} = N_1\left(\dot{\gamma}\right), \quad S_{yy} - S_{zz} = N_2\left(\dot{\gamma}\right), \tag{6.18}$$

respectively. These normal stress differences are even function of the shear rate, and they vanish when the shear rate is zero. This is made explicit by writing

$$N_1\left(\dot{\gamma}\right) = \dot{\gamma}^2\nu_1\left(\dot{\gamma}\right), \quad N_2\left(\dot{\gamma}\right) = \dot{\gamma}^2\nu_2\left(\dot{\gamma}\right), \tag{6.19}$$

where ν_1 and ν_2 are called the first and second normal stress coefficients. They are even functions of the shear rate. Collectively, are called *viscometric functions*. They are material properties.

In a general steady viscometric flow, the above reasoning continues to hold, and we write the stress tensor using the base vectors $\mathbf{a}$, $\mathbf{b}$, $\mathbf{c}$ as

$$\mathbf{T} = -P\mathbf{I} + \dot{\gamma}\eta\left(\mathbf{ab} + \mathbf{ba}\right) + \left(N_1 + N_2\right)\left(\mathbf{aa} + \mathbf{bb}\right) - N_1\mathbf{bb}.$$

Here, P is arbitrary, $\mathbf{I} = \mathbf{aa} + \mathbf{bb} + \mathbf{cc}$ is the unit tensor, and

$$\mathbf{A}_1 = \dot{\gamma}\left(\mathbf{ab} + \mathbf{ba}\right), \quad \mathbf{A}_1^2 = \dot{\gamma}^2\left(\mathbf{aa} + \mathbf{bb}\right), \tag{6.20}$$

$$\mathbf{A}_2 = \mathbf{A}_1\mathbf{L} + \mathbf{L}^T\mathbf{A}_1 = \dot{\gamma}^2\left(\mathbf{ab} + \mathbf{ba}\right)\mathbf{ab} + \dot{\gamma}^2\mathbf{ba}\left(\mathbf{ab} + \mathbf{ba}\right) = 2\dot{\gamma}^2\mathbf{bb}.$$

Omitting the arbitrary pressure, the extra stress can be written as

$$\mathbf{S} = \eta\mathbf{A}_1 + \left(\nu_1 + \nu_2\right)\mathbf{A}_1^2 - \frac{\nu_1}{2}\mathbf{A}_2. \tag{6.21}$$

---

[1] The fluid has no way of knowing that the experimenter has suddenly changed his mind and re-defined x_1 as $-x_1$. It will continue merrily reporting the same shear rate and stress. To the experimenter, however, he will notice that the shear rate and the shear stress have the same magnitudes as before, but they have changed signs. He therefore concludes that the shear stress is an odd function of the shear rate. The same story applies to normal stresses; they are even functions of the shear rate.

This resembles the second-order fluid model (4.71), but with important differences. The second-order fluid model is a slow-flow approximation to the simple fluid, and all the coefficients of the model are constant, whereas (6.21) is a restriction on the simple fluid in steady viscometric flows, and therefore is valid *only* in steady viscometric flows for *all* simple fluids. All the coefficients of (6.21) are functions of the strain rate. Importantly, (6.21) is not a model of a fictitious fluid, but is a proven theorem for steady viscometric flows [31].

6.2.1 Controllable and Partially Controllable Flows

If the velocity field can be determined (with or without inertia), no matter what form the viscometric functions may take, then the flow is said to be *controllable*. There are flows in which the kinematics are fully determined by the viscosity function alone – the normal stress differences do not influence the velocity field. Such flows are called *partially controllable*.

Problems

Problem 6.1 Show that the velocity gradient for (6.3) is

$$\mathbf{L} = \dot{\gamma}\mathbf{ab}. \tag{6.22}$$

Consequently, show that all the flows represented by (6.3) are isochoric. Show that the shear rate is $|\dot{\gamma}|$.

Problem 6.2 Show that the shear rate for the helicoidal flow is

$$\dot{\gamma} = \left(r^2 + c^2\right)\nabla\omega \cdot \nabla\omega. \tag{6.23}$$

Problem 6.3 Consider the shear flow between two tilted plates: one plate is at rest and the other, tilted at an angle θ_0, is moving with a velocity U in the $\mathbf{k}$–direction, as shown in Fig. 6.2. Show that

$$\mathbf{u} = U\frac{\theta}{\theta_0}\mathbf{e}_z. \tag{6.24}$$

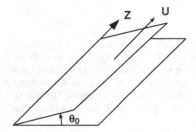

Fig. 6.2. Shear flow between inclined planes

Show that the stress is given by

$$\mathbf{T} = -P\mathbf{I} + \eta\dot\gamma\left(\mathbf{e}_z\mathbf{e}_\theta + \mathbf{e}_\theta\mathbf{e}_z\right) + \left(N_1 + N_2\right)\mathbf{e}_z\mathbf{e}_z + N_2\mathbf{e}_\theta\mathbf{e}_\theta, \qquad (6.25)$$

where the shear rate is $\dot\gamma = U/r\theta_0$, and

$$P = P\left(r_0\right) + I_2\left(\dot\gamma\right) - I_2\left(\dot\gamma_0\right), \quad I_2\left(\dot\gamma\right) = \int_0^{\dot\gamma} \dot\gamma\nu_2 d\dot\gamma. \qquad (6.26)$$

Suggest a way to measure N_2 based on this.

Problem 6.4 The flow between two parallel, coaxial disks is called torsional flow. In this flow, the bottom disk is fixed, and the top disk rotates at an angular velocity of Ω. The distance between the disks is h. Neglecting the fluid inertia, show that

$$\mathbf{u} = \Omega r\frac{z}{h}\mathbf{e}_\theta, \quad \dot\gamma = \Omega\frac{r}{h}. \qquad (6.27)$$

Show that the moment required to turn the disk is

$$M = 2\pi \int_0^R \dot\gamma\eta\left(\dot\gamma\right) r^2 dr, \qquad (6.28)$$

where R is the radius of the disks. Show that the pressure is

$$P\left(r\right) = \int_{\dot\gamma}^{\dot\gamma_0} \dot\gamma\left(\nu_1 + \nu_2\right) d\dot\gamma. \qquad (6.29)$$

From the axial stress, show that the normal force on the top disk is

$$F = \pi R^2 \dot\gamma_0^{-2} \int_0^{\dot\gamma_0} \dot\gamma\left(N_1 - N_2\right) d\dot\gamma. \qquad (6.30)$$

By normalizing the moment and the force as

$$m = \frac{M}{2\pi R^3}, \quad f = \frac{F}{\pi R^2}, \qquad (6.31)$$

show that

$$\eta\left(\dot\gamma_0\right) = \frac{m}{\dot\gamma_0}\left[3 + \frac{d\ln m}{d\ln\dot\gamma_0}\right], \qquad (6.32)$$

and

$$N_1\left(\dot\gamma_0\right) - N_2\left(\dot\gamma_0\right) = f\left(2 + \frac{d\ln f}{d\ln\dot\gamma_0}\right). \qquad (6.33)$$

Relations (6.32–6.33) are the basis for the operation of the parallel-disk viscometer.

Problem 6.5 In a pipe flow, of radius R and pressure drop/unit length $\Delta P/L$, show that the flow rate is

$$Q = 8\pi \left(\frac{L}{\Delta P}\right)^3 \int_0^{\tau_w} \frac{\tau^3}{\eta} d\tau, \tag{6.34}$$

where τ is the shear stress, and τ_w is the shear stress at the wall. In terms of the reduced discharge rate,

$$q = \frac{Q}{\pi R^3}, \tag{6.35}$$

show that

$$\frac{dq}{d\tau_w} = \frac{1}{\eta(\tau_w)} - \frac{3q}{\tau_w}, \tag{6.36}$$

$$\dot{\gamma}_w = q(\tau_w) \left[3 + \frac{d\ln q}{d\ln \tau_w}\right]. \tag{6.37}$$

Formula (6.37) is due to Rabinowitch [32] and is the basis for capillary viscometry.

7. Polymer Solutions

From atoms to flows

In the microstructure approach, a relevant model for the microstructure is postulated, and the consequences are then explored at the macrostructural level, with appropriate averages being taken to smear out the details of the microstructures. The advantage of this is that the resulting constitutive equation is expected to be relevant to the material concerned; and if a particular phenomenon is not well modelled, the microstructural model can be revisited and the relevant physics put in place. This iterative model-building process is always to be preferred over the continuum approach. In this chapter, we will concentrate on the constitutive modelling of dilute polymer solutions.

7.1 Characteristics of a Polymer Chain

Viscoelastic fluids are predominantly long-chain polymer molecules dissolved in a solvent. We will not be concerned with polymer chemistry aspects. It is sufficient for us to know that a typical polymer, with a molecular weight of the order 10^7 grams per mole, has about 10^5 repeating (monomer) units, the monomer molecular weight is of the order 10^2 grams per mole.

7.1.1 Random-Walk Model

The simplest model of a polymer molecule is to replace a real polymer chain with N segments, as illustrated in Fig. 7.1. Each segment has length b, oriented randomly and independently in space. Although the segments are considered to be rigid, they are not physical entities and can cross over each other in space. The end-to-end vector is given by

$$\mathbf{R} = \sum_{j=1}^{N} \mathbf{R}_j. \tag{7.1}$$

On the average, we expect that a randomly oriented vector will have zero mean, and its square is constant:

$$\langle \mathbf{R}_j \rangle = 0, \quad \langle \mathbf{R}_j \cdot \mathbf{R}_j \rangle = b^2 \quad \text{(no sum)}. \tag{7.2}$$

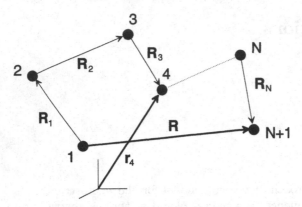

Fig. 7.1. A random-walk model of a polymer chain

Here and elsewhere, the angular brackets denote the average with respect to the probability density function of the variable concerned. Thus, if $P(\mathbf{R}, t)\,d\mathbf{R}$ is the probability of finding a segment of configuration between $\mathbf{R}$ and $\mathbf{R}+d\mathbf{R}$ at time t then the n-th moment of $\mathbf{R}$ is defined as

$$\left\langle \underbrace{\mathbf{R}\mathbf{R}\ldots\mathbf{R}}_{n \text{ times}} \right\rangle = \int \underbrace{\mathbf{R}\mathbf{R}\ldots\mathbf{R}}_{n \text{ times}} P(\mathbf{R}, t)\,d\mathbf{R}. \tag{7.3}$$

Thus, the end-to-end vector has zero mean,

$$\langle \mathbf{R} \rangle = \sum_{j=1}^{N} \langle \mathbf{R}_j \rangle = 0. \tag{7.4}$$

Its mean square is

$$\langle \mathbf{R} \cdot \mathbf{R} \rangle = \langle R^2 \rangle = \sum_{j=1}^{N}\sum_{k=1}^{N} \langle \mathbf{R}_j \cdot \mathbf{R}_k \rangle = \sum_{j=1}^{N} \langle \mathbf{R}_j \cdot \mathbf{R}_j \rangle + \sum_{j=1}^{N}\sum_{k \neq j}^{N} \langle \mathbf{R}_j \cdot \mathbf{R}_k \rangle.$$

Since $\mathbf{R}_j$ is independent of $\mathbf{R}_k$, $k \neq j$, the last double sum is zero. From (7.2), we have

$$\langle R^2 \rangle = Nb^2. \tag{7.5}$$

This model is called the random walk model of a freely rotating chain. Non-freely-rotating chains, for example, chains where the bond angle between successive segments remains fixed, has been considered [33] – they all lead to

$$\langle R^2 \rangle = kNb^2, \tag{7.6}$$

where k is a constant depending on the geometry.

Strong Flow. Thus, in the random walk model, a chain of extended contour length Nb shall have a dimension of $O\left(\sqrt{Nb}\right)$. It is difficult to unravel a polymer molecule, and flows that can do this are called strong.

Diffusion Equation. For a given chain of N segments, the end-to-end vector is a stochastic quantity[1] and must be characterized by its probability density function $P(\mathbf{R}; N)$. Now consider a chain of end-to-end vector with N segments. The probability of $N + 1$ segments having an end-to-end vector $\mathbf{R}$ is precisely the probability of the last segment $(N + 1)$ having a bond vector $\mathbf{b}$, conditional on the first N segments having an end-to-end vector of $\mathbf{R} - \mathbf{b}$:

$$P(\mathbf{R}; N + 1) = \int P(\mathbf{R} - \mathbf{b}; N) P_b(\mathbf{b}) \, d\mathbf{b}. \tag{7.7}$$

This is the property of Markovian processes [34] – namely what happens at any given instant depends only on the instantaneous state of the system, not on its previous history; $P_b(\mathbf{b})$ is called the transition probability, which depends on both current and next states of the process. Here we assume that $\mathbf{b}$ is completely independent of the current state. The equation can be expanded in a Taylor series for $|\mathbf{b}| \ll |\mathbf{R}|$,

$$P(\mathbf{R}; N) + \frac{\partial}{\partial N} P(\mathbf{R}; N) + \cdots$$
$$= \int P_b(\mathbf{b}) \left[P(\mathbf{R}; N) - \mathbf{b} \cdot \nabla P(\mathbf{R}; N) + \frac{1}{2!} \mathbf{b}\mathbf{b} : \nabla\nabla P(\mathbf{R}; N) + \cdots \right] d\mathbf{b}.$$

The average with respect to $\mathbf{b}$ is taken, noting that its distribution is random,

$$\int P_b(\mathbf{b}) \, d\mathbf{b} = 1, \quad \int \mathbf{b} P_b(\mathbf{b}) \, d\mathbf{b} = 0, \quad \int \mathbf{b}\mathbf{b} P_b(\mathbf{b}) \, d\mathbf{b} = \frac{b^2}{3} \mathbf{I}.$$

Thus, we obtain the following diffusion equation for the process $\mathbf{R}$:

$$\frac{\partial}{\partial N} P(\mathbf{R}; N) = \frac{b^2}{6} \nabla^2 P(\mathbf{R}; N). \tag{7.8}$$

The solution of this, subjected to the "initial condition"

$$\lim_{N \to 0} P(\mathbf{R}; N) = \delta(\mathbf{R})$$

is the Gaussian distribution

$$P(\mathbf{R}; N) = \left(\frac{3}{2\pi N b^2} \right)^{3/2} \exp\left(-\frac{3R^2}{2Nb^2} \right). \tag{7.9}$$

This distribution is unrealistic in the sense that there is a finite probability for $R > Nb$; a more exact treatment will produce the *Langevin distribution* [33], which vanishes at $R \geq Nb$ as required.

---

[1] A stochastic process is a family of random variable $X(t)$, where t is the time, X is a random variable, and $X(t)$ is the value observed at time t. The totality of $\{X(t), t \in \mathbb{R}\}$ is said to be a random function or a stochastic process.

7.2 Forces on a Chain

In simple models for dilute polymer solutions, such as the Rouse bead-spring model shown in Fig. 7.2a, a polymer chain is discretized into several effective segments, called *Kuhn segments*,[2] each of which has a point mass (bead) undergoing some motion in a solvent (which is treated as a continuum). Each Kuhn segment may contain several monomer units. Each bead accelerates in response to the forces exerted on it by the solvent, the flow process, and the surrounding beads, and consequently the chain will adopt a configuration. The task here is to relate the microstructure information to a constitutive description of the fluid. When there are only two beads, the model is called the *elastic dumbbell model* [35], Fig. 7.2b, which has been most popular in elucidating the main features of the rheology of dilute polymer solutions.

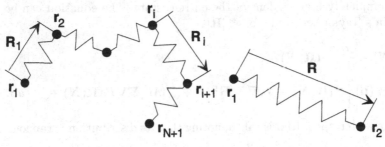

(a) Rouse Bead and Spring Model   (b) Elastic Dumbbell Model

Fig. 7.2. (a) Rouse model and (b) the elastic dumbbell model of a polymer chain

The forces acting on the beads include:

Hydrodynamic forces: These arise from the average hydrodynamic resistance of the motion of the polymer through a viscous solvent. Since the relevant Reynolds number based on the size of the polymer chain is negligibly small, the average motion of the chain is governed by Stokes equations, and Stokes resistance can be used to model this. The chain is usually treated as a number of discrete points of resistance, each having a frictional coefficient. In the simplest model, a frictional force of

$$\mathbf{F}_i^{(d)} = \zeta \left(\mathbf{u}_i - \dot{\mathbf{r}}_i \right), \tag{7.10}$$

is assumed to be acting on the i-th bead, which has a velocity $\dot{\mathbf{r}}_i - \mathbf{u}_i$ relative to the solvent, and ζ is a constant frictional coefficient, which is

---

[2] W. Kuhn (1899–1963) was a Professor at the Technische Hochschule in Karlruhe, and later on, in Basel, Switzerland. He is most famous for the f-summation theorem in quantum mechanics.

usually taken as $6\pi\eta_s a$, where η_s is the viscosity of the solvent, and a represents the size of the bead. To obtain a more realistic model of the nature of the dependence of the frictional forces on the configuration and the deformation of the polymer chain, ζ can be allowed to depend on the length of the segment i, or indeed it may be considered to be a second-order tensor, which reflects the physical idea that the resistance to the motion perpendicular to the chain is much higher than that along the chain. This simple frictional model neglects hydrodynamic interaction with other beads, usually called the *free-draining* assumption; the hydrodynamic interaction arises because of the solvent velocity containing disturbance terms due to the presence of other beads.

Tension in the chain: A chain in equilibrium will tend to curl up into a spherical configuration, with the most probable zero end-to-end vector. However, if the chain ends are forcibly extended, then there is a tension or a spring force, arising in the chain, solely due to the fewer configurations available to the chain. To find the expression for the chain tension, we recall that the probability density function is proportional to the number of configurations available to the chain (i.e., the entropy), and thus the Helmholtz free energy of the chain is [33]

$$F_r\left(\mathbf{r}_j\right) = A\left(T\right) - kT \ln P\left(\mathbf{r}_j\right), \tag{7.11}$$

where $A\left(T\right)$ is a function of the temperature alone. The entropic spring force acting on bead i is

$$\mathbf{F}_i^{(s)} = -\frac{\partial F_r}{\partial \mathbf{r}_i} = kT \frac{\partial \ln P}{\partial \mathbf{r}_i}. \tag{7.12}$$

For the Gaussian chain (7.9), the tension required to extend the chain by a vector $\mathbf{R}$ is

$$\mathbf{F} = \frac{3kT}{Nb^2} \mathbf{R}. \tag{7.13}$$

This applies to individual beads of Fig. 7.2a. Thus if each segment consists of n Kuhn segments, each of bond length b, then the force on bead i due to the chain tension is

$$\mathbf{F}_i^{(s)} = \frac{3kT}{nb^2}\left(\mathbf{r}_{i+1} - \mathbf{r}_i + \mathbf{r}_{i-1} - \mathbf{r}_i\right) = \frac{3kT}{nb^2}\left(\mathbf{R}_i - \mathbf{R}_{i-1}\right). \tag{7.14}$$

This implies that the beads are connected by linear springs of stiffness

$$H = \frac{3kT}{nb^2}. \tag{7.15}$$

A distribution, which better accounts for the finite segment length, is the Langevin distribution,[3] and this results in the so-called inverse Langevin spring law for the chain tension:

---

[3] P. Langevin (1872–1940) introduced the stochastic DE (7.20) in 1908, and showed that the particle obeys the same diffusion equation as described by Einstein (1905).

$$L\left(\frac{bF}{kT}\right) = \coth\left(\frac{bF}{kT}\right) - \frac{bF}{kT} = \frac{r}{nb}, \tag{7.16}$$

where F is the magnitude of the force, r the magnitude of the extension, and the Langevin function is defined as $L(x) = \coth x - x$. A useful approximation of the Langevin spring law is the Warner spring [2]

$$H_i = \frac{3kT}{nb^2}\frac{1}{1 - (R_i/L_i)^2}, \tag{7.17}$$

where $L_i = nb$ is the maximum extended length of segment i. This stiffness approaches infinity as $R_i \to L_i$.

Brownian forces: Brownian forces are the cumulative effect of the bombardment of the chain by the solvent molecules.[4] These forces have a small correlation time scale, typically the vibration period of a solvent molecule, of the order 10^{-13}s for water molecules. If we are interested in time scales considerably larger than this correlation time scale, then the Brownian forces acting on bead i, $\mathbf{F}_i^{(b)}$, can be considered as white noise having a zero mean and a delta autocorrelation function:[5]

$$\left\langle \mathbf{F}_i^{(b)}(t)\right\rangle = 0, \quad \left\langle \mathbf{F}_i^{(b)}(t+s)\,\mathbf{F}_j^{(b)}(t)\right\rangle = 2\delta_{ij}\delta(s)\,\mathbf{f}. \tag{7.18}$$

This states that the strength of the Brownian forces is the measure of the integral correlation function over a time scale, which is considerably greater than the correlation time scale of the Brownian forces:

Fig. 7.3. The French physicist Jean Baptiste Perrin (1870–1942) gave the correct explanation to the random motion of small particles as observed by Brown and confirmed the theoretical calculations by Einstein. For this work he was awarded the Nobel Prize for Physics in 1926. He was also the founder of the Centre National de la Recherche Scientifique

---

[4] The random zig-zag motion of small particles (less than about 10μm) is named after R. Brown (1773–1858), an English botanist, who mistook this as a sign of life. He travelled with Matthew Flinders to Australia in 1801 on the ship *Investigator* as a naturalist. The correct explanation of the phenomenon was given by Perrin (Fig. 7.3). Brownian particles are those undergoing a random walk, or Brownian motion.

[5] This approximation is called white noise, i.e., Gaussian noise of all possible frequencies uniformly distributed. Sometimes it is called "rain-on-the-roof" approximation: two (or more) rain drops do not fall on the same spot on the roof.

$$2\mathbf{f} = \int_{-\infty}^{\infty} \left\langle \mathbf{F}_i^{(b)} \left(t + s \right) \mathbf{F}_i^{(b)} \left(t \right) \right\rangle ds. \tag{7.19}$$

The strength of the Brownian forces is not an arbitrary quantity determined by a constitutive modelling; it is in fact related to the mobility of the Brownian particle.

7.3 Fluctuation-Dissipation Theorem

7.3.1 Langevin Equation

There are several fluctuation-dissipation theorems [36], relating the strength of the fluctuating quantity to the macroscopic "mobility" of the phenomenon concerned. The following development is patterned after Hinch [37].

All micro-mechanical models for a polymer chain in a dilute solution can be written as

$$m\ddot{\mathbf{x}} + \zeta \dot{\mathbf{x}} + \mathbf{K}\mathbf{x} = \mathbf{F}^{(b)} \left(t \right), \tag{7.20}$$

called the *Langevin equation*, where the system state is represented by the finite-dimensional vector $\mathbf{x}$, such that its kinetic energy is $\frac{1}{2}\mathbf{m} : \dot{\mathbf{x}}\dot{\mathbf{x}}$, and its generalized linear momentum is $m\dot{\mathbf{x}}$, $\mathbf{m}$ being a generalized inertia tensor. The inertia tensor m is defined through the kinetic energy, and therefore there is no loss of generality in considering only symmetric $\mathbf{m}$. The system is acted on by a frictional force, which is linear in its state velocities, a restoring force (possibly nonlinear in $\mathbf{x}$), and a Brownian force $\mathbf{F}^{(b)} \left(t \right)$. We assume that the frictional tensor coefficient ζ is symmetric. This system can be conveniently started from rest at time $t = 0$.

Since the Brownian force has only well-defined statistical properties, the Langevin equation (7.20) must be understood as a *stochastic differential equation* [34]. It can only be "solved" by specifying the probability distribution $W \left(\mathbf{u}, \mathbf{x}, t \right)$ of the process $\{\mathbf{u} = \dot{\mathbf{x}}, \mathbf{x}\}$ defined so that $W \left(\mathbf{u}, \mathbf{x}, t \right) d\mathbf{u}d\mathbf{x}$ is the probability of finding the process at the state between $\{\mathbf{u}, \mathbf{x}\}$ and $\{\mathbf{u} + d\mathbf{u}, \mathbf{x} + d\mathbf{x}\}$ at time t. Prescribing the initial conditions $\mathbf{u}(0) = \mathbf{u}_0$, $\mathbf{x}(0) = \mathbf{x}_0$ for (7.20) is equivalent to specifying a delta probability at time $t = 0$:

$$W \left(\mathbf{u}, \mathbf{x}, 0 \right) = \delta \left(\mathbf{u} - \mathbf{u}_0 \right) \delta \left(\mathbf{x} - \mathbf{x}_0 \right).$$

The distribution $W \left(\mathbf{u}, \mathbf{x}, t \right)$ is the *phase space description* of the stochastic process $\{\mathbf{u}, \mathbf{x}\}$. The dependence of W on $\mathbf{x}$, or $\mathbf{u}$ can be eliminated, by integrating out the unwanted independent variable. Then, we have either a *velocity space*, or a *configuration space* description, respectively.

7.3.2 Equi-Partition of Energy

The existence of the temperature T of the surrounding fluid demands that the distribution in the velocity space must satisfy the *equi-partition energy principle*:

$$\lim_{t \to \infty} \langle \dot{\mathbf{x}}(t) \dot{\mathbf{x}}(t) \rangle = kT\mathbf{m}^{-1}, \tag{7.21}$$

as demanded by the kinetic theory of gases, i.e., each mode of vibration is associated with a kinetic energy of $\frac{1}{2}kT$. We now explore the consequence of this on the Langevin system (7.20).

There are three time scales in this system:

1. τ_r the relaxation time scale of the chain in its lowest mode; this time scale is of the order $\left| \zeta \mathbf{K}^{-1} \right|$;
2. τ_i the much shorter inertial relaxation time scale of the chain; this time scale is of the order $\left| \mathbf{m}\zeta^{-1} \right|$;
3. τ_c the still shorter correlation time scale of the Brownian force – this is of the same order as the relaxation time scale of a solvent molecule.

7.3.3 Fluctuation-Dissipation Theorem

In general we have $\tau_c \ll \tau_i \ll \tau_r$, but the estimate of τ_i can vary considerably. To derive a fluctuation-dissipation theorem for the Langevin equation (7.20), we must consider only events at time scale τ_i. In this time scale, $\mathbf{m}(\mathbf{x})$ and $\zeta(\mathbf{x})$ can be replaced by their local values, i.e., regarded as constant, and the state vector can be re-defined to eliminate $\mathbf{Kx}$ so that (7.20) becomes

$$\mathbf{m}\ddot{\mathbf{x}} + \zeta\dot{\mathbf{x}} = \mathbf{F}^{(b)}(t), \tag{7.22}$$

subjected to the initial rest state

$$\dot{\mathbf{x}}(0) = \mathbf{0} = \mathbf{x}(0). \tag{7.23}$$

Note that $\mathbf{A} \cdot e^{\mathbf{A}t} = e^{\mathbf{A}t} \cdot \mathbf{A}$, and

$$\frac{d}{dt}\left(e^{\mathbf{A}t}\mathbf{y}\right) = e^{\mathbf{A}t}\left(\dot{\mathbf{y}} + \mathbf{A}\mathbf{y}\right),$$

the solution is therefore given by

$$\dot{\mathbf{x}}(t) = \int_0^t \exp\left\{\mathbf{m}^{-1}\zeta\left(t' - t\right)\right\} \mathbf{m}^{-1}\mathbf{F}^{(b)}(t')\,dt'. \tag{7.24}$$

This leads to the expectation

$$\langle \dot{\mathbf{x}}(t)\dot{\mathbf{x}}(t) \rangle = \int_0^t \int_0^t \exp\left\{\mathbf{m}^{-1}\zeta\left(t' - t\right)\right\} \mathbf{m}^{-1} \left\langle \mathbf{F}^{(b)}(t') \mathbf{F}^{(b)}(t'') \right\rangle$$
$$\cdot \mathbf{m}^{-1} \exp\left\{\zeta\mathbf{m}^{-1}\left(t'' - t\right)\right\} dt'\,dt''.$$

If $\tau_c \ll \tau_i$, the white noise assumption for the Brownian force can be used, so that

$$\left\langle \mathbf{F}^{(b)}\left(t'\right)\mathbf{F}^{(b)}\left(t''\right)\right\rangle = 2\delta\left(t'-t''\right)\mathbf{f},$$

giving

$$\langle \dot{\mathbf{x}}\left(t\right)\dot{\mathbf{x}}\left(t\right)\rangle = 2\int_0^t \exp\left\{\mathbf{m}^{-1}\boldsymbol{\zeta}\left(t'-t\right)\right\}\mathbf{m}^{-1}\mathbf{f}\mathbf{m}^{-1}\exp\left\{\boldsymbol{\zeta}\mathbf{m}^{-1}\left(t'-t\right)\right\}dt'.$$

This can be integrated by parts to yield

$$\langle \dot{\mathbf{x}}\left(t\right)\dot{\mathbf{x}}\left(t\right)\rangle = -2\exp\left\{-\mathbf{m}^{-1}\boldsymbol{\zeta}\tau\right\}\boldsymbol{\zeta}^{-1}\mathbf{f}\mathbf{m}^{-1}\exp\left\{-\boldsymbol{\zeta}\mathbf{m}^{-1}t\right\}$$
$$+2\boldsymbol{\zeta}^{-1}\mathbf{f}\mathbf{m}^{-1}-\boldsymbol{\zeta}^{-1}\mathbf{m}\langle \dot{\mathbf{x}}\left(t\right)\dot{\mathbf{x}}\left(t\right)\rangle \boldsymbol{\zeta}\mathbf{m}^{-1}.$$

In the limit of $t \to \infty$ (i.e., $t \gg \tau_i$ but $t \ll \tau_r$ so that the equation of state remains linear), the equi-partition of energy (7.21) hold, and we have

$$kT\mathbf{m}^{-1} = 2\boldsymbol{\zeta}^{-1}\mathbf{f}\mathbf{m}^{-1} - kT\mathbf{m}^{-1},$$

or

$$\mathbf{f} = kT\boldsymbol{\zeta}. \qquad (7.25)$$

This is the fluctuation-dissipation theorem, relating the strength of the Brownian force to the mobility of the Brownian system; any dependence on the configuration of $\mathbf{f}$ is inherited from that of $\boldsymbol{\zeta}$.

7.3.4 Diffusivity Stokes–Einstein Relation

A Brownian particle will diffuse in time, this is characterized by the diffusivity

$$\mathbf{D} = \lim_{t \to \infty}\frac{1}{2}\frac{d}{dt}\langle \mathbf{x}\left(t\right)\mathbf{x}\left(t\right)\rangle. \qquad (7.26)$$

This is equivalent to

$$\mathbf{D} = \lim_{t \to \infty}\frac{1}{2}\int_0^t \langle \dot{\mathbf{x}}\left(t\right)\dot{\mathbf{x}}\left(t-\tau\right)+\dot{\mathbf{x}}\left(t-\tau\right)\dot{\mathbf{x}}\left(t\right)\rangle d\tau. \qquad (7.27)$$

7.3.5 Fokker–Planck Equation

As mentioned earlier, the Langevin equation can only be considered solved when the probability function of the process is specified. In the limit $\mathbf{m} \to \mathbf{0}$ it can be shown that the configuration probability density function $\phi\left(\mathbf{x},t\right)$ satisfies

Fig. 7.4. Subrahmanyan Chandrasekhar (1910–1995) was an outstanding Indian astrophysicist. He worked on various aspects of stellar dynamics and was awarded the Nobel Prize in 1983

$$\frac{\partial \phi}{\partial t} = \lim_{\Delta t \to 0} \frac{\partial}{\partial \mathbf{x}} \cdot \left[\frac{\langle \Delta \mathbf{x} \Delta \mathbf{x} \rangle}{2 \Delta t} \cdot \frac{\partial \phi}{\partial \mathbf{x}} - \frac{\langle \Delta \mathbf{x} \rangle}{\Delta t} \phi \right]. \tag{7.28}$$

This is the Fokker–Planck,[6] or Smoluchowski[7] diffusion equation. A clear exposition of this can be found in Chandrasekhar [38], Fig. 7.4.

7.3.6 Smoothed-Out Brownian Force

Since the probability must satisfy

$$\int \phi(\mathbf{x}, t) \, d\mathbf{x} = 1, \tag{7.29}$$

an application of the Reynolds transport theorem yields

$$\frac{\partial \phi}{\partial t} + \frac{\partial}{\partial \mathbf{x}} \cdot (\dot{\mathbf{x}} \phi) = 0, \tag{7.30}$$

assuming that we deal with an equivalent "deterministic" system $\mathbf{x}$. By comparing (7.30) to (7.72), the velocity of this equivalent system must

$$\dot{\mathbf{x}} = -\zeta^{-1} \cdot \mathbf{K} \mathbf{x} - kT \zeta^{-1} \cdot \frac{\partial \ln \phi}{\partial \mathbf{x}}$$

that is,

$$\zeta \dot{\mathbf{x}} + \mathbf{K} \mathbf{x} = -kT \frac{\partial \ln \phi}{\partial \mathbf{x}}. \tag{7.31}$$

Comparing this to (7.69), it is as though the Brownian force has been replaced by

$$\mathbf{F}^{(b)}(t) = -kT \frac{\partial \ln \phi}{\partial \mathbf{x}}. \tag{7.32}$$

---

[6] A.D. Fokker derived the diffusion equation for a Brownian particle in velocity space in 1914. The general case was considered by M. Planck (1858–1947) in 1917.

[7] The general solution to the random walk problem in one dimension was obtained by M. von Smoluchowski in 1906.

Of course, this equation is not mathematically meaningful: the left side is a stochastic, and the right side is a deterministic quantity. It is so defined for the sole purpose of getting the correct diffusion equation (7.72). It is called the *smoothed-out Brownian force*, and is a device employed in most texts dealing with kinetic theories of polymers, e.g., Bird et al. [2].

7.4 Stress Tensor

There are several ways to derive the expression for the stress tensor contributed by the polymer chains in a dilute solution. One is the probabilistic approach (Bird et al. [2]), where the number of polymer chains straddling a surface and the net force acting on that surface by the chains is calculated. The force per unit area can be related to the stress tensor. Another approach is to calculate the free energy of the chain from its entropy, and the rate of work done can be related to the dissipation due to the presence of the chains from which the expression for the stress tensor can be derived [39]. In this section, using a mechanistic approach, treating the chain as a continuum, derives the expression for the stress tensor.

Consider a model for a polymer chain, for example, the bead-spring chain shown in Fig. 7.5. The tension in the i-th Kuhn segment is denoted by $\mathbf{f}_i$. If the chain is Gaussian, then

$$\mathbf{f}_i = H_i \mathbf{R}_i, \text{ (no sum)} \tag{7.33}$$

where $\mathbf{R}_i$ is the end-to-end vector, and H_i is the stiffness of segment i, given in (7.15) – the bead-spring model in this case is called the Rouse model. Using the approach of Landau and Lifshitz [36] and Batchelor [4], the fluid is taken as an effective continuum made up of a homogeneous suspension of

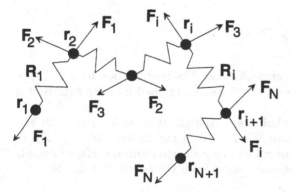

Fig. 7.5. Connector force in a Rouse chain

polymer chains, also regarded as a continuum. Then the effective stress in the fluid is simply the volume-averaged stress:

$$\langle \mathbf{T} \rangle = \frac{1}{V} \int_V \mathbf{T} dV = \frac{1}{V} \int_{V_s} \mathbf{T} dV + \frac{1}{V} \sum_p \int_{V_p} \mathbf{T} dV,$$

where $\mathbf{T}$ is the total stress, V is a representative volume containing several chains, and is made up of a solvent volume V_s and a polymer volume $\sum V_p$. In the solvent volume, the stress is simply the solvent stress, and we have

$$\frac{1}{V} \int_{V_s} \mathbf{T}^{(s)} dV = \frac{1}{V} \int_V \mathbf{T}^{(s)} dV - \frac{1}{V} \sum_p \int_{V_p} \mathbf{T}^{(s)} dV.$$

With a Newtonian solvent, we obtain

$$\frac{1}{V} \int_V \mathbf{T}^{(s)} dV = -p_1 \mathbf{I} + 2\eta_s \mathbf{D}, \tag{7.34}$$

where p_1 is the hydrostatic pressure, η_s the solvent viscosity, and $\mathbf{D}$ the strain rate tensor. The remaining terms are

$$\frac{1}{V} \sum_p \int_{V_p} \mathbf{T}^{(s)} dV = \frac{1}{V} \sum_p \int_{V_p} \left[-p\mathbf{I} + \eta_s \left(\nabla \mathbf{u} + \nabla \mathbf{u}^T \right) \right] dV$$

$$= -p_2 \mathbf{I} + \frac{\eta_s}{V} \sum_p \int_{S_p} (\mathbf{un} + \mathbf{nu}) \, dS. \tag{7.35}$$

Since the chain is modelled as a series of discrete beads, where the interaction with the solvent and other segments takes place, the surface of the chain p consists of the surfaces of the beads. Since the connectors are entirely fictitious, they are allowed to cross one another; thus this chain model is sometimes called the phantom chain.

On the bead surface, the velocity is regarded as uniform, and can be taken out of the integral. Thus

$$\int_{S_p} \mathbf{un} dS = \mathbf{u} \int_{S_p} \mathbf{n} dS = \mathbf{0},$$

by an application of the divergence theorem. The contribution from (7.35) is therefore only an isotropic stress, which can be lumped into the hydrostatic pressure.

Next, if we consider the chain as a continuum as well, then from the force equilibrium we must have $\nabla \cdot \mathbf{T} = \mathbf{0}$ in the chain, and thus $T_{ik} = \partial(T_{ij}x_k)/\partial x_j$. The volume integral can be converted into a surface integral, and the contribution to the effective stress from the polymer chains is

$$\frac{1}{V} \sum_p \int_{V_p} \mathbf{T} dV = \frac{1}{V} \sum_p \int_{S_p} \mathbf{xT} \cdot \mathbf{n} dS = \nu \int_{S_p} \mathbf{xT} \cdot \mathbf{n} dS, \tag{7.36}$$

where S_p is the surface of a representative chain in V, $\mathbf{T} \cdot \mathbf{n}$ is the traction arising in the chain due to the interaction with the flow, and ν is the number density of the chain (number of chains per unit volume); the passage to the second equality is permissible because of the homogeneity assumption which allows us to just consider one generic chain. We can now replace the integral in (7.36) by a sum of integrals over the beads:

$$\int_{S_p} \mathbf{x}\mathbf{T} \cdot \mathbf{n}dS = \sum_i \int_{\text{bead } i} \mathbf{x}\mathbf{T} \cdot \mathbf{n}dS.$$

On bead i, $\mathbf{x}$ can be replaced by $\mathbf{r}_i$ and taken outside the integral, and the remaining integral of the traction on the surface of bead i is therefore the drag force, which bead i exerts on the solvent, and which is proportional to the velocity of the bead relative to that of the solvent:

$$\mathbf{F}_i^{(d)} = -\zeta \left(\dot{\mathbf{r}}_i - \mathbf{u}_i \right).$$

In the absence of inertia, this force is equal to the connector forces plus the Brownian forces acting on the beads:

$$\int_{S_p} \mathbf{x}\mathbf{T} \cdot \mathbf{n}dS = -\mathbf{r}_1\mathbf{f}_1 + \mathbf{r}_2\left(\mathbf{f}_1 - \mathbf{f}_2\right) + \cdots + \mathbf{r}_{N+1}\mathbf{f}_N - \sum_{i=1}^{N+1} \mathbf{r}_i\mathbf{F}_i^{(b)}$$

$$= \sum_{i=1}^{N} \mathbf{R}_i\mathbf{f}_i - \sum_{i=1}^{N+1} \mathbf{r}_i\mathbf{F}_i^{(b)}.$$

Next, the ensemble average with respect to the distribution function of $\mathbf{R}_i$ is taken. The contribution from the Brownian forces is only an isotropic stress, as can be shown either by using the expression for the smoothed-out Brownian forces, or by integrating the Langevin equations directly. Using the expression for the smoothed-out Brownian force,

$$\sum_{i=1}^{N+1} \mathbf{r}_i\mathbf{F}_i^{(b)} = -kT \sum_{i=1}^{N+1} \int \mathbf{r}_i \frac{\partial \ln \phi}{\partial \mathbf{r}_i} \phi d\mathbf{r}_1 \ldots d\mathbf{r}_{N+1}$$

$$= -kT \sum_{i=1}^{N+1} \int \mathbf{r}_i \frac{\partial \phi}{\partial \mathbf{r}_i} d\mathbf{r}_1 \ldots d\mathbf{r}_{N+1}.$$

The integral can be evaluated by parts,

$$\int \mathbf{r}_i \frac{\partial \phi}{\partial \mathbf{r}_i} d\mathbf{r}_1 \ldots d\mathbf{r}_{N+1} = \int \frac{\partial}{\partial \mathbf{r}_i} \left(\mathbf{r}_i \phi \right) d\mathbf{r}_1 \ldots d\mathbf{r}_{N+1} - \mathbf{I} \int \phi d\mathbf{r}_1 \ldots d\mathbf{r}_{N+1}.$$

The first integral on the right is a volume integral over an unbounded domain. It can be converted into surface integral, on the surface at infinity and therefore vanishes, because $\phi \to 0$ there. The second integral on the right is unity, because ϕ is the probability density function. Thus,

$$\sum_{i=1}^{N+1} \mathbf{r}_i \mathbf{F}_i^{(b)} = kT(N+1)\mathbf{I}. \tag{7.37}$$

7.4.1 Kramers Form

All these isotropic stresses can be absorbed in the pressure, and the polymer-contributed stress is

$$\mathbf{S}^{(p)} = \nu \sum_{i=1}^{N} \langle \mathbf{R}_i \mathbf{f}_i \rangle = \nu \sum_{i=1}^{N} \langle H_i \mathbf{R}_i \mathbf{R}_i \rangle, \tag{7.38}$$

which is called the Kramers form for the polymer-contributed stress. The total stress tensor in a dilute polymer solution is

$$\langle \mathbf{T} \rangle = -p\mathbf{I} + \mathbf{S}^{(s)} + \mathbf{S}^{(p)} = -p\mathbf{I} + 2\eta_s \mathbf{D} + \nu \sum_{i=1}^{N} \langle H_i \mathbf{R}_i \mathbf{R}_i \rangle. \tag{7.39}$$

7.5 Elastic Dumbbell Model

The simplest model designed to capture the slowest, and in many ways, the most important relaxation mode of a polymer chain, is the elastic dumbbell model first proposed by Kuhn [35] (Fig. 7.2b). Here we care only about the end-to-end vector of the polymer chain, and all interactions between the solvent and the chain are localized at two beads, located at the chain ends, $\mathbf{r}_1$ and $\mathbf{r}_2$. Each bead is associated with a frictional factor ζ and a negligible mass m. We will assume a Gaussian chain, with the constant spring stiffness

$$H = \frac{3kT}{Nb^2},$$

where N is the number of effective Kuhn segments in the dumbbell, each of which has an extended length b. Furthermore, the frictional coefficient $\zeta = 6\pi\eta_s a$ is assumed to be constant, where η_s is the solvent viscosity and a represents the radius of the beads. The model is also called the *linear elastic dumbbell* model to emphasize the linear force law being used. Although the general equations have been developed in the previous section, it is instructive to write down all the equations again, for this particular case.

7.5.1 Langevin Equations

The equations of motion are,

$$\begin{aligned} m\ddot{\mathbf{r}}_1 &= \zeta(\mathbf{u}_1 - \dot{\mathbf{r}}_1) + H(\mathbf{r}_2 - \mathbf{r}_1) + \mathbf{F}_1^{(b)}(t), \\ m\ddot{\mathbf{r}}_2 &= \zeta(\mathbf{u}_2 - \dot{\mathbf{r}}_2) + H(\mathbf{r}_1 - \mathbf{r}_2) + \mathbf{F}_2^{(b)}(t), \end{aligned} \tag{7.40}$$

where $\mathbf{u}_i = \mathbf{u}(\mathbf{r}_i)$ is the fluid velocity evaluated at the location of the bead i, and $\mathbf{F}_i^{(b)}(t)$ is the Brownian force acting on bead i. The fluctuation-dissipation theorem (7.25) can be used to relate the strength of the Brownian forces to the mobility of the beads:

$$\left\langle \mathbf{F}_j^{(b)}(t) \right\rangle = 0, \quad \left\langle \mathbf{F}_i^{(b)}(t+s)\, \mathbf{F}_j^{(b)}(t) \right\rangle = 2kT\zeta\delta(s)\,\delta_{ij}\mathbf{I}. \tag{7.41}$$

Let us now define the centre of gravity and the end-to-end vector of the dumbbell through

$$\mathbf{R}^{(c)} = \frac{1}{2}(\mathbf{r}_2 + \mathbf{r}_1), \quad \mathbf{R} = \mathbf{r}_2 - \mathbf{r}_1, \tag{7.42}$$

and expand the velocity about the centre of gravity,

$$\begin{aligned}
\mathbf{u}_1 &= \mathbf{u}^{(c)} - \tfrac{1}{2}\mathbf{R}\cdot\nabla\mathbf{u}^{(c)} + \tfrac{1}{8}\mathbf{R}\mathbf{R}:\nabla\nabla\mathbf{u}^{(c)} + O\left(R^3\right), \\
\mathbf{u}_2 &= \mathbf{u}^{(c)} + \tfrac{1}{2}\mathbf{R}\cdot\nabla\mathbf{u}^{(c)} + \tfrac{1}{8}\mathbf{R}\mathbf{R}:\nabla\nabla\mathbf{u}^{(c)} + O\left(R^3\right),
\end{aligned} \tag{7.43}$$

where the superscript c denotes an evaluation at the centre of gravity. From (7.40),

$$\begin{aligned}
m\ddot{\mathbf{R}}^{(c)} &= \zeta\left(\mathbf{u}^{(c)} - \dot{\mathbf{R}}^{(c)}\right) + \tfrac{1}{8}\zeta\mathbf{R}\mathbf{R}:\nabla\nabla\mathbf{u}^{(c)} + \mathbf{F}^{(b,c)}(t), \\
m\ddot{\mathbf{R}} &= \zeta\left(\mathbf{L}\mathbf{R} - \dot{\mathbf{R}}\right) - 2H\mathbf{R} + \mathbf{F}^{(b)}(t),
\end{aligned} \tag{7.44}$$

where $\mathbf{L} = \left(\nabla\mathbf{u}^{(c)}\right)^T$ is the velocity gradient and

$$\mathbf{F}^{(b,c)} = \frac{1}{2}\left(\mathbf{F}_1^{(b)} + \mathbf{F}_2^{(b)}\right), \quad \mathbf{F}^{(b)} = \mathbf{F}_2^{(b)} - \mathbf{F}_1^{(b)} \tag{7.45}$$

are the Brownian forces acting on the centre of gravity and the end-to-end vector. From (7.41),

$$\begin{aligned}
&\left\langle \mathbf{F}^{(b,c)}(t) \right\rangle = 0, \quad \left\langle \mathbf{F}^{(b,c)}(t+s)\,\mathbf{F}^{(b,c)}(t) \right\rangle = kT\zeta\delta(s)\,\mathbf{I}, \\
&\left\langle \mathbf{F}^{(b)}(t) \right\rangle = 0, \quad \left\langle \mathbf{F}^{(b)}(t+s)\,\mathbf{F}^{(b)}(t) \right\rangle = 4kT\zeta\delta(s)\,\mathbf{I}.
\end{aligned} \tag{7.46}$$

With negligible mass, the Langevin equations (7.44) become

$$\begin{aligned}
\dot{\mathbf{R}}^{(c)} &= \mathbf{u}^{(c)} + \tfrac{1}{8}\mathbf{R}\mathbf{R}:\nabla\nabla\mathbf{u}^{(c)} + \zeta^{-1}\mathbf{F}^{(b,c)}(t), \\
\dot{\mathbf{R}} &= \mathbf{L}\mathbf{R} - 2\zeta^{-1}H\mathbf{R} + \zeta^{-1}\mathbf{F}^{(b)}(t).
\end{aligned} \tag{7.47}$$

7.5.2 Average Motion

If the flow is homogeneous, i.e., $\mathbf{L}$ is constant, then the centre of gravity drifts just like a particle of fluid,

$$\left\langle \dot{\mathbf{R}}^{(c)} \right\rangle = \left\langle \mathbf{u}^{(c)} \right\rangle = \mathbf{L}\left\langle \mathbf{R}^{(c)} \right\rangle = \mathbf{u}\left(\left\langle \mathbf{R}^{(c)} \right\rangle\right). \tag{7.48}$$

A migration from the streamline of the centre of gravity will be induced by a heterogeneous flow field. This migration is

$$\left\langle \dot{\mathbf{R}}^{(c)} - \mathbf{u}^{(c)} \right\rangle = \frac{1}{8} \left\langle \mathbf{R}\mathbf{R} \right\rangle : \nabla\nabla\mathbf{u}^{(c)}. \tag{7.49}$$

The average end-to-end vector evolves in time according to

$$\left\langle \dot{\mathbf{R}} \right\rangle = \mathbf{L} \left\langle \mathbf{R} \right\rangle - 2H\zeta^{-1} \left\langle \mathbf{R} \right\rangle, \tag{7.50}$$

which consists of a flow-induced stretching (first term on the right) plus a restoring mechanism (second term on the right) due to the connector spring force. The parameter

$$\lambda = \frac{\zeta}{4H} = \frac{\zeta N b^2}{12kT} \tag{7.51}$$

is called the *Rouse relaxation time*.

7.5.3 Strong and Weak Flows

Equation (7.50) has been used as a basis for delineating between strong and weak flows: strong flows are those in which the flow-induced deformation overcomes the restoring force allowing to grow exponentially in time. Otherwise the flow is weak.

Since we are more interested in the end-to-end vector, the process $\mathbf{R}^{(c)}$ can now be discarded. The Fokker–Planck equation for the density distribution function for $\mathbf{R}$ reads

$$\frac{\partial \phi}{\partial t} = \frac{\partial}{\partial \mathbf{R}} \cdot \left[\frac{2kT}{\zeta} \frac{\partial \phi}{\partial \mathbf{R}} - \left(\mathbf{L}\mathbf{R} - \frac{2H}{\zeta}\mathbf{R} \right) \phi \right]. \tag{7.52}$$

In many cases, there is no need to find the full probability distribution – all we want is the mean, $\left\langle \mathbf{R}\mathbf{R} \right\rangle$, since this is related to the stress, and this can be found without solving for ϕ.

First,

$$\frac{d}{dt}\mathbf{R}\mathbf{R} = \dot{\mathbf{R}}\mathbf{R} + \mathbf{R}\dot{\mathbf{R}}$$

$$= \mathbf{L} \cdot \mathbf{R}\mathbf{R} + \mathbf{R}\mathbf{R} \cdot \mathbf{L}^T - 4H\zeta^{-1}\mathbf{R}\mathbf{R} + \zeta^{-1}\left(\mathbf{R}\mathbf{F}^{(b)} + \mathbf{F}^{(b)}\mathbf{R} \right).$$

Secondly, $\mathbf{R}$ and $\mathbf{F}^{(b)}$ have widely different time scales. Thus

$$\left\langle \mathbf{R}\mathbf{F}^{(b)} \right\rangle = \left\langle \mathbf{R}\left(\Delta t\right)\mathbf{F}^{(b)}\left(\Delta t\right) \right\rangle$$

$$= \left\langle \left(\left[\mathbf{L}\mathbf{R} - 2H\zeta^{-1}\mathbf{R}\right]\Delta t + \zeta^{-1}\int_0^{\Delta t} \mathbf{F}^{(b)}\left(t\right) dt \right) \mathbf{F}^{(b)}\left(\Delta t\right) \right\rangle$$

$$= \zeta^{-1}\int_0^{\Delta t} \left\langle \mathbf{F}^{(b)}\left(t\right)\mathbf{F}^{(b)}\left(\Delta t\right) \right\rangle dt = 4kT\mathbf{I}\int_0^{\Delta t} \delta\left(t - \Delta t\right) dt$$

$$= 2kT\mathbf{I}.$$

Thus

$$\frac{d}{dt}\langle\mathbf{RR}\rangle = \mathbf{L}\langle\mathbf{RR}\rangle + \langle\mathbf{RR}\rangle\mathbf{L}^T - \frac{4H}{\zeta}\langle\mathbf{RR}\rangle + \frac{4kT}{\zeta}\mathbf{I}.$$

Re-arranging, and recalling (7.51),

$$\langle\mathbf{RR}\rangle + \lambda\left\{\frac{d}{dt}\langle\mathbf{RR}\rangle - \mathbf{L}\langle\mathbf{RR}\rangle - \langle\mathbf{RR}\rangle\mathbf{L}^T\right\} = \frac{1}{3}Nb^2\mathbf{I}. \tag{7.53}$$

Next using (7.39),

$$\langle\mathbf{T}\rangle = -p\mathbf{I} + \mathbf{S}^{(s)} + \mathbf{S}^{(p)} = -p\mathbf{I} + 2\eta_s\mathbf{D} + \nu H\langle\mathbf{RR}\rangle. \tag{7.54}$$

7.5.4 Upper-Convected Maxwell Model

Since the polymer-contributed stress is proportional to $\langle\mathbf{RR}\rangle$, it satisfies

$$\mathbf{S}^{(p)} + \lambda\left\{\frac{d}{dt}\mathbf{S}^{(p)} - \mathbf{L}\mathbf{S}^{(p)} - \mathbf{S}^{(p)}\mathbf{L}^T\right\} = G\mathbf{I}, \tag{7.55}$$

where

$$G = \frac{1}{3}Nb^2\nu H = \nu kT. \tag{7.56}$$

The derivative in the braces is called upper convected derivative, one of the many derivatives introduced by Oldroyd [12] to guarantee the stress tensor objectivity. Denoting the upper-convected derivative by $\delta/\delta t$:

$$\frac{\delta\mathbf{A}}{\delta t} = \frac{d\mathbf{A}}{dt} - \mathbf{L}\mathbf{A} - \mathbf{A}\mathbf{L}^T. \tag{7.57}$$

We customary re-define the polymer-contributed stress as

$$\mathbf{S}^{(p)} = G\mathbf{I} + \tau^{(p)}, \tag{7.58}$$

then, since

$$\frac{\delta}{\delta t}\mathbf{I} = -(\mathbf{L} + \mathbf{L}^T) = -2\mathbf{D},$$

we obtain

$$\tau^{(p)} + \lambda\frac{\delta}{\delta t}\tau^{(p)} = 2\eta_p\mathbf{D}, \tag{7.59}$$

where

$$\eta_p = G\lambda = \frac{\nu\zeta Nb^2}{12} = \frac{1}{2}\pi\nu aNb^2\eta_s \tag{7.60}$$

is the polymer-contributed viscosity. The model (7.55), or (7.59), is called the Upper Convected Maxwell (UCM) model in honour of Maxwell, who introduced the linear version in his kinetic theory of gases in 1867. Since $\mathbf{S}^{(p)}$ is proportional to $\langle\mathbf{RR}\rangle$, it is positive definite, whereas $\tau^{(p)}$ is not. In some numerical applications, (7.55) may be preferred to (7.59), because checking the lack of positive definiteness in $\mathbf{S}^{(p)}$ can be easily implemented; this lack can be used as an indication of impending numerical divergence.

7.5.5 Oldroyd-B Model

When the solvent and the polymer-contributed stresses are combined, c.f. (7.54),

$$\mathbf{S} = \mathbf{S}^{(s)} + \tau^{(p)} = 2\eta_s \mathbf{D} + \tau^{(p)}, \tag{7.61}$$

one has

$$(\mathbf{S} - 2\eta_s \mathbf{D}) + \lambda \frac{\delta}{\delta t} (\mathbf{S} - 2\eta_s \mathbf{D}) = 2\eta_p \mathbf{D}$$

$$\mathbf{S} + \lambda_1 \frac{\delta \mathbf{S}}{\delta t} = 2\eta \left(\mathbf{D} + \lambda_2 \frac{\delta \mathbf{D}}{\delta t} \right), \tag{7.62}$$

where $\lambda_1 = \lambda$ is the relaxation time, $\eta = \eta_s + \eta_p$ is the total viscosity, $\lambda_2 = \lambda \eta_s / \eta$ is the retardation time. The model (7.62) is called the Oldroyd fluid B model.

The Oldroyd-B constitutive equation qualitatively describes many features of the so-called Boger fluids.[8] In a steady state simple shear flow, this constitutive equation predicts a constant viscosity, a first normal stress difference which is quadratic in the shear rate, and a zero second normal stress difference. In an unsteady state shear flow, the stresses increase monotonically in time, without overshoot usually observed in some dilute polymer solutions. In an elongational flow, the elongational viscosity becomes infinite at a finite elongation rate of $1/(2\lambda)$ – these will be explored in a series of problems.

7.6 Main Features of the Oldroyd-B Model

Recall the relative strain tensor

$$\mathbf{C}_t (t - s) = \mathbf{F}_t (s)^T \mathbf{F}_t (s), \ \mathbf{C}_t (s)^{-1} = \mathbf{F}_t (s)^{-1} \mathbf{F}_t (s)^{-T},$$

$$\mathbf{F}_t (s) = \mathbf{F} (t - s) \mathbf{F} (t)^{-1}, \ \mathbf{F}_t (t - s)^{-1} = \mathbf{F} (t) \mathbf{F} (t - s)^{-1},$$

$$\dot{\mathbf{F}} (t) = \mathbf{L} (t) \mathbf{F} (t), \ \dot{\mathbf{F}}_t (s)^{-1} = \mathbf{L} (t) \mathbf{F}_t (s), \ \dot{\mathbf{F}}_t (s)^{-T} = \mathbf{F}_t (s) \mathbf{L} (t)^T,$$

$$\dot{\mathbf{C}}_t (s)^{-1} = \mathbf{L} (t) \mathbf{C}_t (s)^{-1} + \mathbf{C}_t (s)^{-1} \mathbf{L} (t)^T. \tag{7.63}$$

7.6.1 Simple Flows

In a simple shear flow, with a time-dependent shear rate $\dot{\gamma} (t)$, the stress components of the UCM model (7.59) obey

---

[8] Dilute solutions of polymers in highly viscous solvents [40].

$$\tau_{11}^{(p)} + \lambda \left\{ \dot{\tau}_{11}^{(p)} - 2\dot{\gamma}\tau_{12}^{(p)} \right\} = 0,$$
$$\tau_{22}^{(p)} + \lambda \dot{\tau}_{22}^{(p)} = 0,$$
$$\tau_{33}^{(p)} + \lambda \dot{\tau}_{33}^{(p)} = 0,$$
$$\tau_{12}^{(p)} + \lambda \left\{ \dot{\tau}_{12}^{(p)} - \dot{\gamma}\tau_{22}^{(p)} \right\} = \eta_p \dot{\gamma}.$$

$$(7.64)$$

If the stresses start from zero initial states, then $\tau_{22}^{(p)} = \tau_{33}^{(p)} = 0$ for all time, and the only two non-trivial components are $\tau_{11}^{(p)}$ and $\tau_{12}^{(p)}$.

In an oscillatory flow with shear rate $\dot{\gamma} = \dot{\gamma}_0 \Re \left(e^{i\omega t} \right)$, where $\Re$ denotes the real part,

$$\tau_{11}^{(p)} + \lambda \dot{\tau}_{11}^{(p)} = 2\dot{\gamma}_0 \Re \left(e^{i\omega t} \right) \tau_{12}^{(p)}, \quad \tau_{12}^{(p)} + \lambda \dot{\tau}_{12}^{(p)} = \eta_p \dot{\gamma}_0 \Re \left(e^{i\omega t} \right). \qquad (7.65)$$

The prediction of an infinite stress at a finite elongational rate is not physically realistic. It is due to the linear dumbbell model being allowed to stretch infinitely. Constraining the dumbbell to a maximum allowable length will fix this problem (e.g., FENE dumbell, Phan-Thien/Tanner model [2] [5] [7]).

The linear elastic dumbbell model is also inadequate in oscillatory flow: it predicts a shear stress proportional to the amplitude of the shear strain, irrespective of the latter magnitude. This is unrealistic: in practice this proportionality is only found when the shear strain is small (less than about 10% for polymer solutions and melts).

7.6.2 Multiple Relaxation Time UCM Model

The frequency response of the dumbell model is also inadequate, due to only one relaxation time in the model. With multiple relaxation times, the Rouse model, Fig. 7.1a, results in

$$\tau^{(p)} = \sum_{j=1}^{N} \tau^{(j)}, \qquad (7.66)$$

$$\tau^{(j)} + \lambda_j \frac{\delta \tau^{(j)}}{\delta t} = 2\eta_j \mathbf{D},$$

where $\{\lambda_j, \eta_j\}$ is the discrete relaxation spectrum. The dynamic properties are now much improved:

$$\eta' = \eta_s + \sum_{j=1}^{N} \frac{\eta_j}{1 + \lambda_j^2 \omega^2}, \quad \eta'' = \sum_{j=1}^{N} \frac{\eta_j \lambda_j \omega}{1 + \lambda_j^2 \omega^2}, \qquad (7.67)$$

$$G' = \sum_{j=1}^{N} \frac{G_j \lambda_j^2 \omega^2}{1 + \lambda_j^2 \omega^2}, \quad G'' = \eta_s \omega + \sum_{j=1}^{N} \frac{G_j \lambda_j \omega}{1 + \lambda_j^2 \omega^2}.$$

In a steady shear flow, the model predicts a constant viscosity, a quadratic first normal stress difference in the shear rate, and a zero second normal stress difference. The Boger fluids show little shear thinning over a large range of shear rates, but this is no doubt due to the high solvent viscosity that completely masks the contribution from the polymer viscosity; any amount of shear-thinning from the polymer contribution would hardly affect on the total fluid viscosity. In general, dilute polymer solutions usually show some degree of shear thinning. The fix is to adopt a more realistic force law for the chain. One such model is the FENE (Finitely Extendable Nonlinear Elastic) model [2].

Problems

Problem 7.1 Use the solution (7.24) in (7.27) to show that

$$\mathbf{D} = kT\zeta^{-1}. \tag{7.68}$$

This is the *Stokes–Einstein relation*, relating the diffusivity to the mobility of a Brownian particle.

Problem 7.2 Starting from the Langevin equation in configuration space, in the limit $\mathbf{m} \to \mathbf{0}$,

$$\dot{\mathbf{x}} = -\zeta^{-1} \cdot \mathbf{Kx} + \zeta^{-1}\mathbf{F}^{(b)}(t), \tag{7.69}$$

show that

$$\Delta\mathbf{x}(t) = -\zeta^{-1} \cdot \mathbf{Kx}\Delta t + \int_t^{t+\Delta t} \zeta^{-1}\mathbf{F}^{(b)}(t')\,dt'. \tag{7.70}$$

From this, show that

$$\langle \Delta\mathbf{x} \rangle = -\zeta^{-1} \cdot \mathbf{Kx}, \quad \langle \Delta\mathbf{x}(t)\,\Delta\mathbf{x}(t) \rangle = 2kT\zeta\Delta t \tag{7.71}$$

and conclude that the Fokker–Planck equation is

$$\frac{\partial\phi}{\partial t} = \frac{\partial}{\partial\mathbf{x}} \cdot \left[kT\zeta^{-1}\frac{\partial\phi}{\partial\mathbf{x}} + \zeta^{-1} \cdot \mathbf{Kx}\phi \right]. \tag{7.72}$$

Problem 7.3 Investigate the migration problem in a plane Poiseuille flow.

Problem 7.4 Show that the solution to (7.50) is

$$\langle \mathbf{R}(t) \rangle = e^{-t/2\lambda}e^{\mathbf{L}t}\mathbf{R}_0. \tag{7.73}$$

Thus conclude that the flow is strong if

$$\text{eigen}(\mathbf{L}) \geq 1/2\lambda,$$

where eigen $(\mathbf{L})$ is the maximum eigenvalue of $\mathbf{L}$.

Problem 7.5 Using the result (7.63), show that the following solves the Maxwell equation (7.55):

$$\mathbf{S}^{(p)}(t) = \frac{G}{\lambda} \int_{-\infty}^{t} e^{(s-t)/\lambda} \mathbf{C}_t(s)^{-1} \, ds = G\mathbf{I} + \tau^{(p)}. \tag{7.74}$$

This integral version of the UCM model is called the Lodge rubber-like liquid model [41]. It was derived from a network of "rubber" strands, a model meant for concentrated polymer solutions and melts. It is remarkable that two distinct microstructures, a dilute suspension of dumbbells and a concentrated network of polymer strands, share a common constitutive framework.

8. Suspensions

Particulates

Suspension is a term used to describe an effective fluid made up of parcles suspended in a liquid; examples of such liquids abound in natural and man-made materials: blood, mild, paints, inks. The concept of a suspension is meaningful only when there are two widely different length scales in the problem: l is a typical dimension of a suspended particle and L is a typical size of the apparatus, and $l \ll L$. When this is not met, we have a collection of discrete individual particles suspended in a liquid. Most progress has been made with Newtonian suspensions, i.e., suspensions of particles in a Newtonian liquid. The review paper by Metzner [42] contained most of the relevant information on the subject.

If the particles are small enough (less than 10μm in size), then they will undergo Brownian motion, and the micromechanics is described by a set of stochastic differential equations, together with some relevant fluctuation-dissipation theorems, and the full solution of the relevant equations can only be obtained by specifying the probability distribution of the system, through solving a diffusion equation.

The relative importance of Brownian motion is characterised by a Péclet number, such as $Pe = O(\eta_s \dot{\gamma} l^3 / kT)$, the ratio of viscous stress to stress induced by thermal excitation, where $\dot{\gamma}$ is a typical strain rate, η_s is the solvent viscosity, and kT is the Boltzmann temperature. At low Péclet numbers, Brownian motion is strong, and the particles' orientation tends to be randomised, leading to a larger dissipation (i.e., higher effective viscosity) than when the Péclet number is large, the Brownian motion is weak, and the particles tend to align with the flow most of the time. Thus, we expect shear-thinning with the inclusion of Brownian motion (increasing shear rate leads to an increase in the Péclet number). We will focus on non-Brownian flow regime, where the particles are large enough ($\gtrsim 10\mu$m in size), but yet orders of magnitudes smaller than L.

With $l \ll L$, the microscale Reynolds number is small. Thus, the relevant equations governing the micromechanics are the Stokes equations,

$$\nabla \cdot \mathbf{u} = 0, \quad -\nabla p + \eta \nabla^2 \mathbf{u} = \mathbf{0}. \tag{8.1}$$

Stokes equations are linear and instantaneous in the driving boundary data. Consequently the microdynamics are also linear and instantaneous in the driving forces; only the present boundary data are important, not their past history. This does not imply that the overall response will have no memory, nor does it imply that the macroscaled Reynolds number is small. In most studies, the particle's inertia is neglected, its inclusion may lead to a non-objective constitutive model, since the stress contributed by micro inertia may not be objective (Ryskin and Rallison [16]).

The linearity of the micromechanics implies that the particle-contributed stress will be linear in the strain rate; in particular all the rheological properties (shear stress, first and second normal stress differences, and elongational stresses) will be linear in the strain rate. Several investigators have indeed found Newtonian behaviour in shear for suspensions up to a large volume fraction. However, experiments with some concentrated suspensions usually show shear-thinning behaviour, but the particles in these experiments are in the μm range, where Brownian motion would be important. Shear-thickening behaviour, and indeed, yield stress and discontinuous behaviour in the viscosity-shear-rate relation have been observed, e.g., Metzner [42]. This behaviour cannot be accommodated within the framework of hydrodynamic interaction alone; for a structure to be formed, we need forces and torques of a non-hydrodynamic origin.

8.1 Bulk Suspension Properties

Consider now a volume V which is large enough to contain many particles but small enough so that the macroscopic variables hardly change on the scale $V^{1/3}$, i.e., $l \ll V^{1/3} \ll L$. The effective stress tensor seen from a macroscopic level is simply be the volume-averaged stress [36] [4],

$$\langle \sigma_{ij} \rangle = \frac{1}{V} \int_V \sigma_{ij} \, dV = \frac{1}{V} \int_{V_f} \sigma_{ij} \, dV + \frac{1}{V} \int_{V_p} \sigma_{ij} \, dV,$$

where V_f is the volume occupied by the solvent, V_p is the volume of the particles in V, and the angle brackets denote a volume-averaged quantity. If the solvent is Newtonian, we have

$$\sigma_{ij}(\mathbf{x}) = -p\delta_{ij} + 2\eta_s D_{ij} \equiv \sigma_{ij}^{(f)}, \quad \mathbf{x} \in V_f.$$

Thus

$$\frac{1}{V} \int_{V_f} \sigma_{ij} dV = -\langle p \rangle \, \delta_{ij} + 2\eta_s \langle D_{ij} \rangle - \frac{1}{V} \int_{V_p} \sigma_{ij}^{(f)} dV.$$

Furthermore, from the equations of motion in the absence of inertia and the body force,

$$\frac{\partial}{\partial x_k} (x_i \sigma_{kj}) = \sigma_{ij},$$

and we find that

$$\frac{1}{V} \int_{V_p} \sigma_{ij} \, dV = \frac{1}{V} \int_{V_p} \frac{\partial}{\partial x_k} (x_i \sigma_{kj}) \, dV = \frac{1}{V} \int_{S_p} x_i t_j \, dS,$$

where $t_j = \sigma_{jk} n_k$ is the traction vector and S_p is the bounding surface of all the particles. In addition,

$$\frac{1}{V} \int_{V_p} \sigma_{ij}^{(f)} \, dV = \frac{1}{V} \int_{V_p} \left[-p\delta_{ij} + \eta_s \left(\frac{\partial u_i}{\partial x_j} + \frac{\partial u_j}{\partial x_i} \right) \right] \, dV$$

$$= \frac{1}{V} \int_{S_p} [\eta_s (u_i n_j + u_j n_i)] \, dS$$

to within an isotropic tensor which can be lumped into a generic hydrostatic pressure P, which is determined through the balance of momentum and the incompressibility constraint. The average stress is thus given by

$$\langle \sigma_{ij} \rangle = \underbrace{-p'\delta_{ij} + 2\eta_s \langle D_{ij} \rangle}_{\text{solvent}} + \underbrace{\frac{1}{V} \int_{S_p} \{x_i t_j - \eta_s (u_i n_j + u_j n_i)\} \, dS}_{\text{particles}}, \quad (8.2)$$

consisting of a solvent contribution, and a particle contribution; p' is just a scalar pressure (the prime will be dropped from hereon). The particle contribution can be decomposed into a symmetric part, and an antisymmetric part. The symmetric part is in fact the sum of the stresslets $S_{ij}^{(p)}$ defined by

$$S_{ij} = \frac{1}{2} \int_{S_p} \{x_i t_j + x_j t_i - 2\eta_s (u_i n_j + u_i n_j)\} \, dS = \sum_p S_{ij}^{(p)}, \quad (8.3)$$

and the antisymmetric part leads to the rotlet:

$$\mathcal{R}_{ij} = \frac{1}{2} \int_{S_p} (x_i t_j - x_j t_i) \, dS = \frac{1}{2} \sum_p \epsilon_{ijk} T_k^{(p)}, \quad (8.4)$$

where $T_k^{(p)}$ is the torque exerted on the particle p, and the summation is over all particles in the volume V. The particle-contributed stress is therefore given by

$$\sigma_{ij}^{(p)} = \frac{1}{V} \sum_p \left(S_{ij}^{(p)} + \frac{1}{2} \epsilon_{ijk} T_k^{(p)} \right). \quad (8.5)$$

The rate of energy dissipation in a large enough volume V to contain all particles is given by

$$\Phi = \int_V \sigma_{ij} D_{ij} \, dV = \int_V \frac{\partial}{\partial x_j} (\sigma_{ij} u_i) \, dV = \int_{S_p} \sigma_{ij} u_i n_j \, dS. \quad (8.6)$$

The second equality comes from the balance of momentum, and the third one from an application of the divergence theorem, assuming that the condition at infinity is quiescent. For a system of rigid particles, the boundary condition on the surface of a particle p is that

$$\mathbf{u} = \mathbf{U}^{(p)} + \Omega^{(p)} \times \mathbf{x}, \tag{8.7}$$

where $\mathbf{U}^{(p)}$ and $\Omega^{(p)}$ are the translational and rotation velocities of the particle, which can be taken outside the integral in (8.6). The terms remaining can be identified with the force $\mathbf{F}^{(p)}$, and the torque $\mathbf{T}^{(p)}$ imparted by the particle to the fluid. Thus the total rate of energy dissipation is

$$\Phi = \sum_p \left(\mathbf{U}^{(p)} \cdot \mathbf{F}^{(p)} + \Omega^{(p)} \cdot \mathbf{T}^{(p)} \right). \tag{8.8}$$

Note also that for a system of rigid particles, the integral

$$\int_{S_p} (\mathbf{un} + \mathbf{nu}) \, dS = 0,$$

since

$$\int_{S_p} \mathbf{U}^{(p)} \mathbf{n} \, dS = 0, \qquad \int_{S_p} \left(\Omega^{(p)} \times \mathbf{xn} + \mathbf{n} \Omega^{(p)} \times \mathbf{x} \right) dS = 0,$$

by applications of the divergence theorem.

8.2 Dilute Suspension of Spheroids

We consider now a dilute suspension of force- and torque-free monodispersed spheres in a general homogeneous deformation. The dilute assumption means the volume fraction

$$\phi = \nu \frac{4\pi a^3}{3} \ll 1, \tag{8.9}$$

where ν is the number density of the spheres, each of radius a. In this case, in a representaive volume V we expect to find only one sphere. Thus, the microscale problem consists of a single sphere in an effectively unbounded fluid; the superscript p on the generic particle can be omitted, and the coordinate system can be conveniently placed at the origin of the sphere. The boundary conditions for this microscale problem are

$$\mathbf{u} = \mathbf{U} + (\mathbf{D} + \mathbf{W}) \cdot \mathbf{x}, \quad \text{far from the particle, } |\mathbf{x}| \to \infty, \tag{8.10}$$

and

$$\mathbf{u} = \mathbf{V} + \mathbf{w} \cdot \mathbf{x}, \quad \text{on the particle's surface, } |\mathbf{x}| = a, \tag{8.11}$$

where $\mathbf{L} = \mathbf{D} + \mathbf{W}$ is the far-field velocity gradient tensor; $\mathbf{D}$ is the strain rate tensor, $\mathbf{W}$ is the vorticity tensor, $\mathbf{w}$ is the skew-symmetric tensor such that $w_{ij} = -\epsilon_{ijk}\Omega_k$, with Ω being the angular velocity of the particle. The far-field boundary condition must be interpreted to be far away from the particle under consideration, but not far enough so that another sphere can be expected. The solution to this unbounded flow problem is well known, [44]

$$\mathbf{u} = \mathbf{U} + \mathbf{L} \cdot \mathbf{x} + \frac{a^3}{x^3}(\mathbf{w} - \mathbf{W}) \cdot \mathbf{x} + \left(\frac{3a}{4x} + \frac{a^3}{4x^3} \right)(\mathbf{V} - \mathbf{U}) - \frac{a^5}{x^5}\mathbf{D} \cdot \mathbf{x} \tag{8.12}$$

$$+ \frac{3(\mathbf{V} - \mathbf{U}) \cdot \mathbf{x}}{x^2}\left(\frac{a}{x} - \frac{a^3}{x^3} \right)\mathbf{x} - \frac{5\mathbf{D} : \mathbf{xx}}{2x^2}\left(\frac{a^3}{x^3} - \frac{a^5}{x^5} \right)\mathbf{x}, \tag{8.13}$$

and

$$p = \frac{3}{2}\eta_s a \frac{(\mathbf{V} - \mathbf{U}) \cdot \mathbf{x}}{x^3} - 5\eta_s a^3 \frac{\mathbf{D} : \mathbf{xx}}{x^5}. \tag{8.14}$$

The traction on the surface of the sphere is

$$\mathbf{t} = \sigma \cdot \mathbf{n}|_{x=a} = -\frac{3\eta_s}{2a}(\mathbf{V} - \mathbf{U}) - \frac{3\eta_s}{a}(\mathbf{w} - \mathbf{W}) \cdot \mathbf{x} + 5\frac{\eta_s}{a}\mathbf{D} \cdot \mathbf{x}. \tag{8.15}$$

The force and the torque on the particle can be evaluated:

$$\mathbf{F} = \int_S \sigma \cdot \mathbf{n} \, dS = -6\pi\eta_s a(\mathbf{V} - \mathbf{U}) \tag{8.16}$$

and

$$\mathbf{T} = \int_S \mathbf{x} \times \sigma \cdot \mathbf{n}dS = -8\pi\eta_s a^3(\Omega - \omega), \tag{8.17}$$

where $\omega_i = \frac{1}{2}\epsilon_{ijk}W_{jk}$ is the local vorticity vector. Thus, if the particle is force-free and torque-free, then it will translate with $\mathbf{U}$ and spin with an angular velocity of ω.

Returning now to the particle-contributed stress, (8.5),

$$\left\langle \sigma_{ij}^{(p)} \right\rangle = \frac{1}{V}\sum_p S_{ij} = \nu S_{ij},$$

where the stresslet is given, from (8.12), and noting that

$$\int_S \mathbf{x} \, dS = 0, \quad \int_S \mathbf{xx} \, dS = \frac{4\pi a^4}{3}\mathbf{1}.$$

we find

$$S_{ij} = \frac{1}{2} \int_S (x_i t_j + x_j t_i) \; dS = 5\frac{\eta_s}{a} \int_S D_{ik} x_k x_j \; dS = 5\eta_s \left(\frac{4\pi a^3}{3}\right) D_{ij},$$

and when we recall that $\phi = 4\pi a^3 \nu / 3$, the effective stress will now become

$$\langle \sigma \rangle = -p\mathbf{1} + 2\eta_s \left(1 + \frac{5}{2}\phi\right)\mathbf{D}. \tag{8.18}$$

This is the celebrated Einstein's result [45] , who arrived at the conclusion from the equality of the dissipation at the microscale and the dissipation at the macroscale as described by an effective Newtonian viscosity.

A similar theory has been worked out for a dilute suspension of spheroids by Leal and Hinch [46] , when the spheroids may be under the influence of Brownian motion, using the solution for flow around a spheroid due to Jeffery [47]. Here, if $\mathbf{p}$ denotes a unit vector directed along the major axis of the spheroid, then Jeffery's solution states that

$$\dot{\mathbf{p}} = \mathbf{W} \cdot \mathbf{p} + \frac{R^2 - 1}{R^2 + 1}(\mathbf{D} \cdot \mathbf{p} - \mathbf{D} : \mathbf{ppp}), \tag{8.19}$$

where R is the aspect ratio of the particle (length to diameter ratio). The particle-contributed stress may be shown to be

$$\sigma^{(p)} = 2\eta_s \phi \{ A\mathbf{D} : \langle \mathbf{pppp} \rangle + B\left(\mathbf{D} \cdot \langle \mathbf{pp} \rangle + \langle \mathbf{pp} \rangle . \mathbf{D}\right) \tag{8.20}$$
$$+ C\mathbf{D} + d_R F \langle \mathbf{pp} \rangle \},$$

where the angular brackets denote the ensemble average with respect to the distribution function of $\mathbf{p}$; A, B, C and F are some shape factors, and d_R is the rotational diffusivity. If the particles are large enough so that Brownian motion can be ignored, then the last term, as well as the angular brackets, can be omitted in the previous expression. The asymptotic values of the shape factors are given in Table 8.1.

Table 8.1. Asymptotic values of the shape factors

Asymptotic limit	$R \to \infty$ (rod-like)	$R = 1 + \delta, \; \delta \ll 1$ (near-sphere)	$R \to 0$ (disk-like)
A	$\dfrac{R^2}{2(\ln 2R - 1.5)}$	$\dfrac{395}{147}\delta^2$	$\dfrac{10}{3\pi R} + \dfrac{208}{9\pi^2} - 2$
B	$\dfrac{6\ln 2R - 11}{R^2}$	$\dfrac{15}{14}\delta - \dfrac{395}{588}\delta^2$	$-\dfrac{8}{3\pi R} + 1 - \dfrac{128}{9\pi^2}$
C	2	$\dfrac{5}{2}\left(1 - \dfrac{2}{7}\delta + \dfrac{1}{3}\delta^2\right)$	$\dfrac{8}{3\pi R}$
F	$\dfrac{3R^2}{\ln 2R - 1/2}$	9δ	$-\dfrac{12}{\pi R}$

The rheological predictions of this constitutive equation have also been considered by Hinch and Leal [48]. In essence, the shear viscosity is shear-thinning, the first normal stress difference is positive while the second normal stress difference is negative, but of a smaller magnitude. The precise values depend on the aspect ratio and the strength of the Brownian motion.

Strictly speaking, the dilute assumption means that the volume fraction is low enough, so that a particle can rotate freely without any hindrance from its nearby neighbours. The distance Δ between any two particles must therefore satisfy $l < \Delta$, so that a volume of l^3 contains only one particle, where l is the length of the particle and d is its diameter. The volume fraction therefore satisfies

$$\phi \sim \frac{d^2 l}{\Delta^3}, \quad \phi R^2 < 1.$$

Thus, the reduced elongational viscosity is only $O(1)$ in the dilute limit, not $O(R^2)$ as suggested by the formula. As the concentration increases, we get subsequently into the semi-dilute regime, the isotropic concentrated solution, and the liquid crystalline solution. The reader is referred to Doi and Edwards [49] for more details. Here, we simply note that the concentration region $1 < \phi R^2 < R$ is called semi-concentrated. Finally, the suspension with $\phi R > 1$ is called concentrated, where the average distance between fibres is less than a fibre diameter, and therefore fibres cannot rotate independently except around their symmetry axes. Any motion of the fibre must necessarily involve a cooperative motion of surrounding fibres. This area is intensively researched on.

8.3 Epilogue

All the cases studied in this books, the microstructure models for polymer solutions and suspensions, set the stage for a constitutive analysis on more complex microstructured materials, and provide the tools for a better understanding of published literature on the subject. We have not mentioned the reptation concept of Doi and Edwards [49], the modelling effort in fibre suspensions [50], in biological materials [51], in electro-rheological fluids, as well as transition phenomena.

The constant theme emphasised throughout is that a relevant evolution equation for the microstructure be derived from well-established physics, together with relevant statistical mechanics linking the microstructure evolution to a macroscaled stress induced by the microstructure. Of course, this stress modifies the global kinematics through the balance equations, which feed back into the microtructure evolution thus completing the cycle. Having a relevant constitutive relation is only half of the story; one needs to be able to make predictions with it, and that usually means a numerical implementation, a vast open area that we have not even mentioned in this book.

It is hoped that the readers find the book useful.

Problems

Problem 8.1 Use the instantaneous nature of the micromechanics to explain the shear reversal experiments of Gadala-Maria and Acrivos [43]. They have found that if shearing is stopped after a steady state has been reached in a Couette device, the torque is reduced to zero instantaneously. If shearing is resumed in the same direction after a period of rest, then the torque would attain its final value that corresponds to the resumed shear rate almost instantaneously. However, if shearing is resumed in the opposite direction, then the torque attains an intermediate value and gradually settles down to a steady state. How would you classify the memory of the liquid, zero, fading or infinite?

Problem 8.2 Show that (8.19) is solved by

$$\mathbf{p} = \frac{\mathbf{Q}}{Q},$$
(8.21)

where

$$\dot{\mathbf{Q}} = \mathcal{L} \cdot \mathbf{Q}, \quad \mathcal{L} = \mathbf{L} - \frac{2}{R^2 + 1}\mathbf{D}.$$
(8.22)

The effective velocity gradient tensor is $\mathcal{L} = \mathbf{L} - \zeta\mathbf{D}$, where $\zeta = 2/(R^2 + 1)$ is a 'non-affine' parameter.

Problem 8.3 In the start-up of a simple shear flow, where the shear rate is $\dot{\gamma}$, show that

$$Q_1 = Q_{10}\cos\omega t + \sqrt{\frac{2-\zeta}{\zeta}}Q_{20}\sin\omega t,$$

$$Q_2 = Q_{20}\cos\omega t - \sqrt{\frac{\zeta}{2-\zeta}}Q_{10}\sin\omega t,$$

and $Q_3 = Q_{30}$, where $\{Q_{10}, Q_{20}, Q_{30}\}$ are the initial components of $\mathbf{Q}$, and the frequency of the oscillation is

$$\omega = \frac{1}{2}\dot{\gamma}\sqrt{\zeta(2-\zeta)} = \frac{\dot{\gamma}R}{R^2+1}.$$

From these results, obtain the particle-contributed stress and the viscometric functions as
The reduced viscosity:

$$\frac{\langle\sigma_{12}\rangle - \eta_s\dot{\gamma}}{\eta_s\dot{\gamma}\phi} = 2Ap_1^2p_2^2 + B\left(p_1^2 + p_2^2\right) + C,$$
(8.23)

The reduced first normal stress difference:

$$\frac{N_1}{\eta_s \dot{\gamma} \phi} = 2Ap_1 p_2 \left(p_1^2 - p_2^2\right),\tag{8.24}$$

and the reduced second normal stress difference:

$$\frac{N_2}{\eta_s \dot{\gamma} \phi} = 2p_1 p_2 \left(Ap_2^2 + B\right).\tag{8.25}$$

Thus, the particles tumble along with the flow, with a period of $T = 2\pi(R^2 + 1)/\dot{\gamma}R$, spending most of their time aligned with the flow.

Problem 8.4 In the start-up of an elongational flow with a positive elongational rate $\dot{\gamma}$, show that

$$Q_1 = Q_{10} \exp\left\{(1 - \varsigma)\dot{\gamma}t\right\},$$
$$Q_2 = Q_{20} \exp\left\{-\frac{1}{2}(1 - \varsigma)\dot{\gamma}t\right\},$$
$$Q_3 = Q_{30} \exp\left\{-\frac{1}{2}(1 - \varsigma)\dot{\gamma}t\right\},$$

so that the particle is quickly aligned with the flow in a time scale $O(\dot{\gamma}^{-1})$. At a steady state, show that the reduced elongational viscosity is given by

$$\frac{N_1 - 3\eta_s \dot{\gamma}}{\eta_s \dot{\gamma} \phi} = 2\left(A + 2B + C\right) \approx \frac{R^2}{\ln 2R - 1.5}.\tag{8.26}$$

References

1. M.E. Gurtin: *An Introduction to Continuum Mechanics.* (Academic Press, New York 1981)
2. R.B. Bird, R.C. Armstrong and O. Hassager: *Dynamics of Polymeric Liquids. Vol 1: Fluid Mechanics,* 2nd Edition. (John Wiley & Sons, New York 1987)
3. O.D. Kellogg: *Foundations of Potential Theory.* (Dover, New York 1954)
4. G.K. Batchelor: J. Fluid Mech. **44** 545–570 (1970)
5. R.I. Tanner: *Engineering Rheology,* Revised Edition. (Oxford University Press, Oxford 1992)
6. R.I. Tanner: J. Polym. Sci. A2, **8** 2067–2078 (1970)
7. R.R. Huilgol and N. Phan-Thien: *Fluid Mechanics of Viscoelasticity: General Principles, Constitutive Modelling, Analytical and Numerical Techniques.* (Elsevier, Amsterdam 1997).
8. F. Brauer and J.A. Nohel: *Qualitative Theory of Ordinary Differential Equations.* (Dover, New York 1989)
9. R.S. Rivlin and J.L. Ericksen: J. Rational Mechanics and Analyses **4** 323–425 (1955)
10. A.E.H. Love: *A Treatise in the Mathematical Theory of Elasticity,* Note B. (Dover, New York 1944)
11. A.C. Pipkin and R.I. Tanner: in *Mechanics Today,* **1** 262 321 Ed S. Nemat-Nasser. (Pergamon, New York 1972)
12. J.G. Oldroyd: Proc. Roy. Soc. Lond. A, **200** 523–541 (1950)
13. R.I. Tanner and K. Walters: *Rheology: An Historical Perspective.* (Elsevier, Amsterdam 1998).
14. W. Noll: J. Ratl. Mech. Anal., **4** 2–81 (1955) ; Arch.
15. W. Noll: Ratl. Mech. Anal., **2** 197–226 (1958)
16. G. Ryskin and J.M. Rallison: J. Fluid Mech., **99** 513–529 (1980)
17. H. Weyl: *Classical Groups.* (Princeton Univ Press, New Jersey 1946)
18. A.C. Pipkin and R.S. Rivlin: Arch. Rational Mech. Anal., **4** 129–144 (1959)
19. A.J.M. Spencer: in *Continuum Physics,* **1**, Ed. Eringen, A.C. (Academic Press, New York, 1971)
20. E. Fahy and G.F. Smith: J. Non-Newt. Fluid Mech., **7**, 33–43 (1980)
21. M. Mooney: J. Applied Phys., **11** 582–592 (1940)
22. L.R.G. Treloar: *The Physics of Rubber Elasticity,* 3rd Edition. (Clarendon Press, Oxford 1975)
23. H. Giesekus: Rheol. Acta, **3** 59–81 (1963)
24. A.C. Pipkin: *Lectures on Viscoelasticity Theory,* 2nd Ed. (Springer-Verlag, Berlin 1986)
25. R.I. Tanner: Phys. Fluids, **9** 1246–1247 (1966)
26. A.E. Green and R.S. Rivlin: Arch. Rat. Mech. Anal., **1** 1–21 (1957)
27. L. Weese: Comp. Phys. Comm., **77** 429–440 (1993)

28. J.L. Ericksen: in *Viscoelasticity - Phenomenological Aspects*, Ed. J.T. Bergen (Academic Press, New York 1960)
29. B.D. Coleman: Arch. Rat. Mech. Anal., **9** 273–300 (1962)
30. W.-L. Yin and A.C. Pipkin: Arch. Rat. Mech. Anal., **37** 111–135 (1970)
31. W.O. Criminale, Jr., J.L. Ericksen and G.L. Filbey, Jr.: Arch. Rat. Mech. Anal., **1** 410–417 (1957)
32. B. Rabinowitch: Z. Physik. Chem. A, **145** 1–26 (1929)
33. P.J. Flory: *Statistical Mechanics of Chain Molecules*. (Wiley Interscience, New York 1969)
34. Y.K. Lin: *Probabilistic Theory of Structural Dynamics*. (McGraw-Hill, New York 1967)
35. W. Kuhn: Kolloid-Zeit., 68 2–15 (1934)
36. L.D. Landau and E.M. Lifshitz: *Fluid Mechanics*, translated by J.B. Sykes and W.H. Reid. (Pergamon Press, New York 1959)
37. E.J. Hinch: J. Fluid Mech., **72** 499–511 (1975)
38. S. Chandrasekhar: Rev. Mod. Phys., **15** 1–89 (1943)
39. R.G. Larson: *Constitutive Equations for Polymer Melts and Solutions*. (Butterworth Publishers, Boston 1988).
40. D.V. Boger and R. Binnington: Trans. Soc. Rheol., **21** 515–534 (1977)
41. A.S. Lodge: *Elastic Liquids*. (Academic Press, New York 1964)
42. A.B. Metzner: J. Rheol., **29** 739–775 (1985)
43. F. Gadala-Maria and A. Acrivos: J. Rheol., **24** 799–814 (1980)
44. J. Happel and H. Brenner: *Low Reynolds Number Hydrodynamics*. (Noordhoff International Publishing, Leyden 1973)
45. A. Einstein: in *Investigations on the Theory of the Brownian Movement*, Ed. R. Fürth, R., Transl. A.D. Cowper. (Dover, New York, 1956)
46. L.G. Leal and E.J. Hinch: Rheol. Acta, **12**, 127–132 (1973)
47. G.B. Jeffery: Proc. Roy. Soc. Lond., A**102**, 161–179 (1922)
48. E.J. Hinch and L.G. Leal: J. Fluid Mech., **52**, 683–712 (1972)
49. M. Doi and S.F. Edwards: *The Theory of Polymer Dynamics*. (Oxford University Press, Oxford, 1988)
50. F.P. Folgar and C.L. Tucker, Jr: J Reinforced Plastics and Composites, **3** 98–119 (1984)
51. E.B. Bagley: *Food Extrusion Science and Technology*. (Marcel Dekker, New York 1992)
52. H. See: J. Phys. D Appl. Phys., **33** 1625–1633 (2000)

Index